The Growing Years

A Study Guide for the Televised Course

The Growing Years

A Study Guide for the Televised Course

Philip Kaushall
Department of Psychology
University of California, San Diego

Kiki Skagen
National Media Office
University Extension
University of California, San Diego

McGraw-Hill Book Company

New York   St. Louis   San Francisco   Auckland   Bogotá
Düsseldorf   Johannesburg   London   Madrid   Mexico
Montreal   New Delhi   Panama   Paris   São Paulo
Singapore   Sydney   Tokyo   Toronto

Library of Congress Cataloging in Publication Data

Kaushall, Philip.
  The growing years.

  "Written as an integrative element for the television programs and the text. *A child's world*."
  Includes index.
  1. Child development.   I. Skagen, Kiki, joint author.
II. Title.   III. A child's world.
RJ131.K38   155.4   77-8308
ISBN   0-07-033455-2

Chapter Opening Photographs

1	Gordon Converse, The Christian Science Monitor	15	Ken Heyman
2	Costa Manos, Magnum	16	Burk Uzzle, Magnum
3	Charles Harbutt, Magnum	17	Burk Uzzle, Magnum
4	Joanne Leonard	18	Eileen Christelow, Jeroboam
5	Suzanne Arms, Jeroboam	19	Erich Hartmann, Magnum
6	Ken Heyman	20	A Staff Photographer, The Christian Science Monitor
7	Linda Rogers, Woodfin Camp	21	Burt Glinn, Magnum
8	Roger Malloch, Magnum	22	Margaretta Mitchell
9	Suzanne Arms, Jeroboam	23	Barbara Klutinis, Jeroboam
10	Costa Manos, Magnum	24	Elizabeth Crews, Jeroboam
11	Russell Abraham, Jeroboam	25	Leo Hetzel
12	Mark Godfrey, Magnum	26	Ken Heyman
13	Elizabeth Crews, Jeroboam	27	Charles Harbutt, Magnum
14	Jeffrey Blankfort, BBM Associates	28	Joanne Leonard
		29	David Powers, Jeroboam
		30	A Staff Photographer, The Christian Science Monitor

The Growing Years
A Study Guide for the Televised Course

1234567890 DODO 783210987

This book was set in Bodoni Book by Holmes Composition Service. The editors were Robert G.
Manley and Holmes Composition Service: the designer was Design Office/Bruce Kortebein: the
production supervisor was Robert C. Pedersen. R. R. Donnelley & Sons Company was the printer
and binder.

Contents

Preface

This study guide is the last component in a process of course development that began in 1975 with discussions among representatives from the University Extension, University of California, San Diego; Coast Community College District; and McGraw-Hill Book Company. Gradually the idea of a television course concerning child development crystallized into a series of meetings involving University Extension and Community College District staff members and academic consultants in the fields of psychology and education. Together these individuals worked out a design for the course and formulated objectives for each unit. Content outlines for the television programs were then developed, reviewed, and the programs went into production. Finally, as scripts and programs became available, *The Growing Years: A Study Guide for the Televised Course* was written as an integrative element for the television programs and the text, *A Child's World*.

Many people have made very substantial contributions to this process. Much of the credit for the initiation of this course belongs to Martin N. Chamberlain and Mary Lindenstein Walshok from the University Extension, University of California, San Diego, Robert Manley of McGraw-Hill Book Company, and Bernard J. Luskin of Coastline Community College. The original academic design team consisted of Jean Mandler, Ph.D., Associate Professor of Psychology at the University of California, San Diego, Astrid Beigel, Ph.D., Senior Psychologist and Program Supervisor, Los Angeles County, Southeast Mental Health Services, Diane Papalia, Ph.D., Associate Professor of Child Development at the University of Wisconsin and coauthor of the text used in the course, and Glenda Riddick, M.A., Instructor of Human Development at Orange Coast College. At a later date Glenda

Riddick was joined by Mary Weir, Ph.D., to provide advice on course content to the television product team.

Many of these people have provided more specific assistance in the preparation of this book. Glenda Riddick and Mary Weir of Long Beach City College have, with Leslie Purdy of Coastline Community College, consulted with the authors throughout its development. Jean Mandler and Astrid Beigel responded to requests for information and advice at all hours. Bernard Scheier of Holmes Composition Service extracted the manuscript from the authors and turned it into a finished product in a remarkably short time. He was aided in this process by Kay Y. James in picture acquisition and Suzanne Knott who did the copy editing. Bruce Kortebein of Design Office was responsible for book design and Toni Tortorilla of University Extension, University of California, San Diego, compiled the glossary.

The authors wish to thank each of these individuals for their assistance as well as the many other people who have participated in the making of the television course of which this book is a part.

Philip Kaushall
Kiki Skagen

The Growing Years

A Study Guide for the Televised Course

1

Introduction

Assignments for this unit

1. Read Overview 1 in this book.
2. View Program 1.
3. Think about your childhood.

Overview 1

This unit is as much an introduction to television courses as it is to child development. Television courses offer students unique opportunities in nontraditional education, but they require a slightly different approach from traditional college courses. We will outline some of these special aspects and then talk about the components of this particular course.

Those of you who have taken television courses before will know that if they are well done, they can offer several kinds of benefits. Students who are unable to attend on-campus classes can receive credit for study done in the home or at off-campus centers. The medium of television permits learning "at a distance," utilizing visual aids for maximum instructional advantage. Through television labs can be visited and experiments demonstrated or children can be viewed in their natural setting while an eminent psychologist interprets their behavior. In addition, the structure of the course allows the individual institution to adapt it to the needs of its own students, adding or subtracting assignments, requiring a lesser or greater number of on-campus sessions, and assigning the amount and type of credit each feels appropriate.

A television course also requires a set of responsibilities that differ from those of the usual on-campus course. It is often not necessary to attend two or three class sessions a week, but it is important to watch each television program for the programs help pace your work and are a basic component of the course. We suggest that you do not try to take notes at all during the programs—the special effectiveness of television is visual, and if your eyes are on your notebook during the program, you will miss this visual component. You may want to jot down important points from the program immediately after it is over for review.

In a television course you may not have a professor telling you the importance of keeping up with your assignments. Different students have different styles of studying, but we feel that the do-no-work-until-just-before-the-exam approach is particularly inappropriate for this type of course. You will gain more from each television program if you have read the assigned pages in *A Child's World*. Furthermore, since this subject is one that builds on knowledge acquired in each unit, you may find you simply do not understand material if you have not done the work for an earlier assignment. In this particular course it is especially important to keep up on the work in the first few units, as they have lengthy reading assignments. You will find that the amount of material to be read decreases after Unit 6, although in many cases this decrease is offset by an increase in the difficulty of the material to be mastered.

The course *The Growing Years* has three basic components: (1) the television programs, (2) the text *A Child's World*, and (3) this study guide. The text carries the bulk of the material you will be expected to master. The television programs amplify and illustrate aspects of each lesson. The study guide integrates the other two elements and adds information that may not have been covered as thoroughly as the course designers felt desirable. Used together, these three components should enable you to fulfill the following course objectives by the end of the 30 units:

1. Recognize genetic, biological, environmental, and sociocultural influences on development.

2. Identify and discuss some of the most important aspects of normal physical, mental, and personality development.

3. Relate child development research and theory to real life situations.

4. Recognize the diversity of approaches in child development research and theory.

5. Distinguish between the popular conceptions of rigid developmental stages and the actual flexibility that occurs in the developmental process.

6. Demonstrate openness and objectivity toward issues, theories, and orientations in child development.

The organization and integration of this material varies slightly from unit to unit, depending on the needs perceived by the course designers. In each lesson some topics may be covered more thoroughly in one component than in the other components. The assignments section indicates where important topics are expanded or omitted.

We have divided child development into sections that make sense to us and, we hope, will make sense to you. However, we do not want you to think that this is the only possible way to approach the subject. Learning is not simple mastery of facts in a particular order. It is the mastery and conceptual organization of information in such a way that you understand it and are able to apply it to different situations. Use the categories, divisions, and linkages we have made in the material, or make your own. In the end the only organization of any material that will carry real meaning to you, the student, is the one you understand.

Similarly, the course designers have had to make decisions about what to include or emphasize and what to omit or treat quickly. Child development is a dynamic field. New material of great importance is continually made available, and there are wide areas of disagreement concerning the interpretation of almost everything children do. Throughout this course you should remember that in most instances there is much more material than has been possible to include.

Those of you who have had psychology courses before will find a number of familiar theories applied to child development. We have attempted to present alternative points of view in most areas. Since these often spring from divergent approaches to psychology in general, we have also tried to provide brief characterizations of Freudian psychology, behaviorism, and other underlying theoretical positions. Not all aspects of child development, however, are treated by each theoretical approach presented. Unit 9 ("The Emerging Personality"), for instance, will present a number of theoretical approaches to personality development. Some of them spring from the same theoretical perspective as approaches to language development or cognitive development; others do not. Names of psychologists associated with theories of development, such as Jean Piaget, may appear in many chapters; others appear in only one. In addition, most of these theories overlap in their approach. You will find three major approaches to child development, however, which are dealt with repeatedly throughout the course: Piagetian, behaviorist, and Freudian/psychoanalytic.

Components of the Study Guide

Each chapter of the study guide is divided into several sections and subsections.

Student Objectives

These outline the basic information you should have mastered by the end of each unit.

Assignments for the Unit

Some of the assignments are standard from unit to unit—the television program and the overview, for example—but others vary. Note carefully the page numbers you are to read in the text and whether or not the last section in the study guide chapter is optional or required.

Overview

This section provides a brief review of most, but not necessarily all, of the material covered in the unit. It also may include some new material or extend portions that have not been covered at great length elsewhere.

Review Questions

These questions are to provide you with a measure of your mastery of the material. Most are not difficult. If you can look through them and feel you know or can find the answers without difficulty, you are probably having no problems with the subjects covered in the unit. If, on the other hand, you glance at them and think, "I could never answer those questions," you may be encountering problems and need to contact your teacher or course facilitator. At the very least, you should read the material again.

We do advise you to answer the questions to be sure that you can. You should also be aware that, although these questions might be useful in reviewing the material for an examination, being able to answer them will not guarantee your being able to answer all the exam questions.

Questions Pondered by Psychologists

These are questions without answers. They are some of the questions that psychologists and others ask as they plan experiments and studies. We have included them because we feel that you should be aware of some of the areas that particularly concern people who are currently working in this field.

Of Policy Matters and Public Interest

Included in this section are controversial topics relating to child development that you may encounter in everyday life. Although we have opinions on most of the topics, we have attempted to be neutral in their presentation. We do not expect the few sentences on these topics to change your attitudes, but we do hope that you will recognize the existence of differing points of view as we do.

2

Studying Children

Student objectives

1. Understand the scope of the field of child development.

2. Identify reasons for studying normal child development.

3. Discuss some of the several ways of conducting research with children. Discuss ethical considerations involved in conducting research involving children as subjects.

4. Recognize some of the basic recurring concepts in the study of child development: critical periods, principal of individual differences, sociocultural influences, range of "normal behavior."

Assignments for this unit

1. Read the Introduction (pages 1–29) in *A Child's World*.

2. View Program 2, "Studying Children."

3. Read Overview 2 in this book and review the study aids.

We are fascinating individuals. Few of us go through a week without thinking of ourselves in one way or another. "I like independence in my work." "My mother really messed me up when I was a kid." "I guess I'm the type people like to talk to." Often the self-analysis is cast in psychological terms, for most of us have at least a passing acquaintance with psychology, however misleading the results of this acquaintance may be. Furthermore, our thoughts about ourselves often turn to our childhoods, for childhood has formed the people we are today. Folk wisdom expresses this in the saying: "The tree grows as the twig is bent."

Child development is concerned with the substance of that maxim. What *is* the effect of "bending the twig?" What exactly are the needs, events, and occurrences that constitute an adult? What is the nature of their mutual influence? What? How? When? Why? The only question missing is who. Who is all of us. Child development may have answers for these and other philosophical questions about human nature. Are humans by nature combative, aggressive, altruistic, or gentle? Are they a result of their heredity or their environment? Are the differences we observe in the behavior of men and women a result of genetics or of the way in which they are reared?

Although these questions are philosophical, our approach to them will be primarily through psychology. Some elements of sociology, anthropology, and education will enter at various points. Regardless of the discipline we use, the field is a relatively new one because this concern with the development of the child did not always exist. The historical development of the concept of *childhood* summarized in the text was paralleled by a development of the study of the individual. Not all eras have been concerned with the individual being in the particular way that ours is. For centuries Western thought looked at the nature of man primarily in relation to God or to the State. Only incidentally was the study of human nature perceived as an exploration toward the fullest realization of human potential, either in this world or the next.

STUDYING CHILDREN

The study of humans (for we include women, an approach that many earlier philosophers and theologians did not take) as a scientific endeavor also followed the evolution of the scientific method. Isaac Newton's spectacular successes in unifying diverse natural facts into a single theory encouraged philosophers to apply Newton's assumptions to the study of human nature. In the seventeenth century John Locke and René Descartes both introduced views that have remained influential to the present day. Locke proposed an empirical theory of children's learning and growth based on environmental stimulation. (Locke's theory is discussed at greater length in Unit 7.) Descartes was convinced that infants were born possessing many complex ideas. He argued that an infant born without any knowledge whatsoever, as Locke contended, would not be able to organize the sensations received from the environment in a coherent manner.

Somewhat later, Jean-Jacques Rousseau proposed the theory that children were like "noble savages" and would develop best if left to do so naturally. Some of this approach is reflected in the followers of Maria Montessori, the Italian physician and educator, and the Montessori preschools. Similarly, Charles Darwin influenced modern psychologists through his naturalistic approach to the study of children. In this approach children are studied through observation of their natural behavior rather than through carefully manipulated experiments.

The study of psychology and child development took a dramatic turn at the turn of the last century. Russian physiologist I. P. Pavlov discovered what is now called the *conditioned reflex* (Unit 7), and Sigmund Freud developed his psychoanalytic approach to the human mind (Unit 9). Both founded schools of thought that remain influential in the study of psychology to this day.

The theories brought to the study of children and the ways in which children are studied both stem from the beliefs of these and other philosophers and psychologists. The orientation of a particular psychologist is important. A researcher's choice of methodology is determined to a very large extent by the type of question studied and the theoretical viewpoint adopted. The nature of the scientific method, however, is one reason why different theories can coexist in psychology; it is not easy to perform a definitive experiment that rules out a competing view since often no single study will do equal justice to the various alternative approaches.

The experiment typifies the scientific method of collecting information, and it is important to understand its logic since it is the source of most scientific data today. In the simplest terms an experiment is designed to answer a specific question. (For example, can training in a bouncing chair accelerate walking ability?) This specific question is usually related to a broader theoretical question. (Can training accelerate a maturational schedule of motor development?) The specific question looks at a *single variable* (bouncing-chain training) and its effect on another variable (walking). The former variable is manipulated by the experiment in one situation and not manipulated in another situation (two groups of infants, one has training, the other group does not). Finally, both variables can be measured, which means that they can be increased or decreased during the experiment without changing qualitatively into something else.

The last point is important. You could not, for example, study the effects of temperature on the size of silkworms if the worms changed into moths halfway through the experiment. In many instances qualitative changes are not very obvious; not taking them properly into account can, as can errors in other aspects of experiments, lead to unjustifiable and unreliable results.

Another requirement that is difficult to follow in practice is that only the variables being studied should change. If other variables change, their effects on the *dependent* variable (in our bouncing-chain case, walking) will be unknown and the results uninterpretable.

Another research method common in the social sciences is the *correlational method*. This is used when the experimenter cannot manipulate the relevant (*independent*—in our case the bouncing-chair training) variable. Instead she is confined to merely observing and measuring the two variables as she finds them. An example might be the influence of climate on intelligence. Intelligence is the dependent variable; it would be hard for an experimenter to change the independent variable either by altering climatic patterns or moving subjects to a different area. You should be aware that the existence of an association is all that the correlational method will reveal. Climate may be associated with certain levels of intelligence (in fact, it is not), but we cannot say the variations in intelligence are *caused* by different climates without other kinds of studies. Other associated variables might be the causative ones.

In the study of child development, *naturalistic observation* is often an important tool for gathering data. Many significant facts can be discovered without statistical techniques or the measurement of variables. Scientists often combine naturalistic observation with other techniques. It may often be used as a preliminary to a controlled experiment. Another combination is observation and manipulation of the situation. A very young infant may not provide an observer with a great deal of spontaneous behavior, so an experimenter might provide it with special toys or sounds and observe its reactions.

In order to study developmental changes from one age to another, scientists can study children repeatedly over the course of months or years (longitudinal studies) or many children in one age group can be compared to a large number of children in another age group (cross-sectional studies). A combination of both longitudinal and cross-sectional methods can be used at the same time in order to provide a greater check on deviations from developmental norms in a particular group.

Throughout this course we will be citing studies using these methods and the various approaches to the subject of child development. We will also find recurring themes that will be discussed in many of the units:

1. We are concerned with the study of *normal* child development because most of us are "normal" people, but the range allowed by that word is great. Normal does not mean

average, and it certainly does not mean *alike*. Joey may run very fast and Archie might be pretty slow, but neither runs *abnormally*.

2. As this example implies, there is a range of differences in each individual. No two people are exactly alike.

3. Many of the differences noted by psychologists in the development of children spring from sociocultural influences. "Sociocultural influences" is a broad topic and has many aspects including cultural and ethnic background, education, religion, and location. We are not sure of exactly what effects most of these aspects have, but we do know that they are important.

4. Most individuals, however, do share some things in common. These include certain sequences of growth and critical periods of growth. Children are able to do things (lift the head, sit, crawl, walk) in a certain order even if these appear at different times. Critical periods are those times when an event will have its greatest impact on development.

Study aids

Review Questions

1. Give two reasons for the study of normal child development.

2. What relation does Pavlov have to the study of child development?

3. A scientist took careful notes and found that children who liked carrots usually had curly hair. He decided that carrots contained a substance that made hair curly and spent the rest of his life looking for it.
    a. What would you call the research method of his initial study?

b. Do you intend to eat more carrots on the basis of this experiment? Why or why not?

4. "When Baby Laura started on solid food, she began eating twice as much."
   a. What was the *qualitative* change?
   b. What was the *quantitative* change?

5. A psychologist wants to study the relationship of how much rats eat to their rate of growth.
   a. What is the *independent* variable?
   b. What is the *dependent* variable?
   c. Is this more apt to be a longitudinal or a cross-sectional study? Explain your answer.
   d. Would there be any effect on the experiment if the experimental group were white rats and the control group giant Australian wharf rats?

Questions Pondered by Psychologists and Other Scientists

How "scientific" can we be in a field such as child development? How much can be accounted for through scientific experiments? How do we control all elements? How do we relate a particular experiment to the whole field of development? Finally, what role does intuition play in science?

Of Policy Matters and Public Interest

To what extent should parents, schools, and others act on the results of psychological experiments? Although some things have been generally agreed upon, newspapers and magazines often report the result of a particular experiment that is still questioned by other psychologists. If it sounds helpful should we accept it or should we wait until there is "more data?" Is there ever enough data? The case of Cyril Burt, discussed in Unit 3 ("Heredity and Environment"), illustrates the needs and perils of such decisions.

3

Heredity and Environment

Student objectives

1.   Discuss the significance of the interaction between heredity and environment on physical, mental, and personality development.

2.   Discuss normal genetic transmission and briefly explain its significance with respect to dominant, recessive, and sex-linked traits.

3.   Cite some examples of genetic abnormalities.

4.   Identify the significance of genotype and phenotype.

Assignments for this unit

1.   Read Chapter 2 (pages 67–97) in *A Child's World*.

2.   View Program 3, "Heredity and Environment."

3.   Read Overview 3 in this book and review the study aids. This is a difficult and highly technical unit. Close attention should be paid to both text and study guide in order to complete the objectives.

4. Read Nature, Nurture, and Intelligence in this chapter. Note that it updates material in the text.

Overview 3

Psychology cannot be separated from biology and chemistry. However, physical development and condition interact throughout life with mental and personality development, and the roots of all of them lie embedded in the molecular structure of *DNA*.

DNA or *deoxyribonucleic acid* is a complex, double-spiral molecule. It occurs in the nuclei of all living cells, can assume a multitude of different forms, and carries genetic information. DNA is also the basic material of *genes*. The composition of DNA in a gene plays a major role in the determination of inherited characteristics for it can alter the effect of the gene. Each gene, in turn, has a specific function to perform in the myriad of traits, growth patterns, and physical characteristics that are passed from generation to generation.

A segment of about 20,000 genes comprises a *chromosome*. Each gene occupies a specific place in the line that forms the chromosome. Furthermore, each gene normally forms a pair with another gene located on a partner chromosome. Normal cells contain 46 individual rod-shaped chromosomes or 23 pairs.

Cells can reproduce themselves into two separate cells through a process of division. In most cases of cell division, or *mitosis*, each offspring has 46 chromosomes or 23 pairs, but in the cells for sexual reproduction, these pairs are split. The cell that results from this form of division or *meiosis* has 23 chromosomes.

Each parent contributes one of these 23 chromosome cells to its child. When the sperm fertilizes the egg, a new cell with the full complement of 46 chromosomes is formed. The structure of the DNA in each of the genes that comes from each parent contributes to the shape and substance of that child. The possible combinations of genes from this process are legion—with the exception of identical twins, the chance of two individuals' having the same genetic make-up is virtually zero.

The actual way genes interact and come together to form most aspects of inheritance is complex and fascinating. Some of the principles, however, can be illustrated using traits that are determined in relatively simple fashion. The genetic code behind the inheritance of blue versus brown eyes is one of these.

One gene pair influences eye color. If the two genes are the same or *homozygous*, the child's eyes will be that color. In other words, if each of Susan's parents contributes a "brown" gene, the child will have brown eyes. If each contributes a "blue" gene, the child's eyes will be blue. What happens if Susan inherits one blue gene and one brown?

The characteristics of a gene pair do not mix. If each of a pair is different, one will be *dominant*. Since the "brown" gene is dominant, Susan's eyes will be brown, but she will have the capacity to pass either a "brown" gene or a "blue" gene to her child. The "blue" gene is *recessive,* which means that blue eyes will not be *expressed* in a boy or girl if there is a "brown" gene present. If Jason has blue eyes, it means that both the genes affecting eye color are "blue." It also means that when Jason grows up, the only eye-color gene he can pass to his offspring will, again, be "blue."

Genotype describes the actual genes carried by an individual. Susan's *genotype* for eye color is of a person carrying one "blue" and one "brown" gene. *Phenotype* is the way in which these genes are expressed. Susan's *phenotype* is that of a brown-eyed person. Nancy, who has two "brown" genes, has the same phenotype as Susan but a different genotype.

The difference between genotype and phenotype is interesting when it concerns the colors of eyes; it becomes much more important when it concerns genetically transmitted disease. If a disease such as sickle-cell anemia is transmitted through a recessive gene, a person who may appear to be free of the disease can, in actuality, be a carrier. If he or she has a child by another carrier and through the process of meiosis and fertilization the two recessive genes are brought together, the child will be born with the disease.

Some of these diseases are linked not only to recessive genes but to the sex of the child. Sex itself is determined by the chromosome "donated" to the child by the father. A woman has two X chromosomes; a man has one X and one Y chromosome. If a man contributes an X chromosome, the child will be a girl; if the chromosome is a Y, a boy will be born. The gene for a sex-linked disease is carried on an X chromosome and is *recessive* if in a person with two X chromosomes—a woman—but becomes dominant in a person who has an X and a Y—a man. A woman, then, can carry a disease such as hemophilia, but only her sons stand a chance of getting it. A fuller explanation of genetically transmitted diseases and of diseases caused by abnormalities in the genes themselves appears in pages 87–89 of the text.

In most cases, the interaction of genes in heredity is not nearly this clear-cut. Not only can chromosome pairs be split in any way during meiosis, but there is also some crossing over of genes; a portion of one chromosome can break off and exchange places with a part of the other member of the pair. This is particularly important because most traits are not the result of one pair of genes. They can spring from a combination of pairs of genes, and the way those genes act can, in turn, be influenced by still other genes. The result is a marvelously random selection of characteristics.

From the moment of fertilization, environment also begins to interact with genetic factors. Heredity and environment do not simply affect the development of a child in varying proportions. There is, rather, a continuous process of mutual influence. Some psychologists describe this by saying that a child may be born with a certain capacity to develop a particular set of characteristics, but how these are expressed depends on environmental influences in combination with other inherited traits. A child may be born with the potential to grow to a certain height. Whether this potential is realized depends on such environmental factors as nutrition. These factors, in turn, can relate to genetically determined factors such as PKU, which inhibits the body's ability to metabolize certain kinds of foods. The well-known nature-or-nurture debate becomes nature *and* nurture. Nature, nurture, and their interaction will be discussed throughout the course for they provide the framework of human development.

Study aids

Review Questions

1. What is the nature/nurture controversy? How does it relate to child development?

2. Jean has blue eyes.
   a. What is her genotype?
   b, What is her phenotype?
   c. If she were a soap opera heroine, married a blue-eyed man, and had a brown-eyed child, what might be in the scriptwriter's imagination?

3. Why are identical twins used in the study of heredity? What are some of the problems involved in their use?

4. If a woman over 35 years of age wonders whether her unborn baby has Down's syndrome, how might she find out?

5. Hemophilia is an example of:
   a. A sex-linked disease
   b. An autosomal dominant disease
   c. An autosomal recessive disease
   d. An environmental disease
   e. None of the above

6. Arrange the following according to physical size:
   a. chromosome
   b. DNA
   c. sperm
   d. gene

7. Describe how heredity and environment may interact in the case of schizophrenia.

8. What recent evidence is there that IQ test results may not be as great an indicator of inherited intelligence as some psychologists have thought?

Questions Pondered by Psychologists and Geneticists

1. Is genetic transmission truly random or is there a "design" in the evolution of a species?

2. Do advances in medicine and public health lead to genetic deterioration of the species through allowing the "unfit" to survive?

3. Is it possible to consider cultural factors as "inherited," thus distinguishing the hereditary endowment of humans from that of animals?

Of Policy Issues and Public Interest

Scientists are learning how to change the structure of DNA, thus changing the genetic information transmitted through heredity. Recombinant DNA has been hailed by some as a means of eliminating certain kinds of diseases and disabilities, of increasing agricultural yields, and of improving the quality of human life. Others point to its threats through deliberate misuse or through accidents in which viruses become deadly killers. The entire issue poses an ethical question of human control over future generations.

Nature, Nurture, and Intelligence

Humans are, in general, poor material for research on heredity. Unlike fruit flies and mice—two of the most common subjects for this kind of work—they take a long time to produce and a longer time to grow to the point where they can again reproduce themselves. More importantly, it is, and we hope will remain, ethically unacceptable to experiment with the breeding of human subjects.

One of the few ways in which scientists have been able to study the relative influences of heredity and environment in human beings is through the study of identical twins. Since both twins have the same genetic makeup, it should be possible to separate heredity from the influences of environment, particularly if we can study identical twins who are separated at birth.

The most extensive and thorough twin studies were conducted through several decades by British psychologist Cyril Burt. In this series of studies Burt found ample material to support his thesis that intelligence was determined primarily through inheritance. His conclusions were widely used in discussion, teaching, and policy making. The examination system of British schools, in which children were given a test at the age of 11 to determine whether they should be tracked for university education (generally some 7%, primarily from the upper classes, eventually went to universities), was influenced in part by Burt's conclusions. Arthur Jensen's argument that the IQ score difference between black and white children in the United States was largely a result of hereditary differences rested, in large measure, upon the information generated by Burt's studies.

In the early 1970s, however, some psychologists began to question Burt's methods. Burt, who died in 1972, had not clearly documented his work; the two people with whom he had worked most closely were elusive at best and, at worst, nonexistent for much of the period covered by the studies.

The story concerning Burt's research methodology became front page news in 1976, shocking the psychological establishment throughout the world. Now, although there is considerable controversy as to whether Burt deliberately falsified his information, most psychologists and scientists accept the fact that Burt's data are no longer to be considered reliable. Those who, on the basis of this work, had estimated that 80% of intelligence was inherited and 20% determined by environment, have, in many cases, revised their estimates. Those who feel it is impossible to separate heredity and environment in the determination of intelligence feel themselves vindicated.

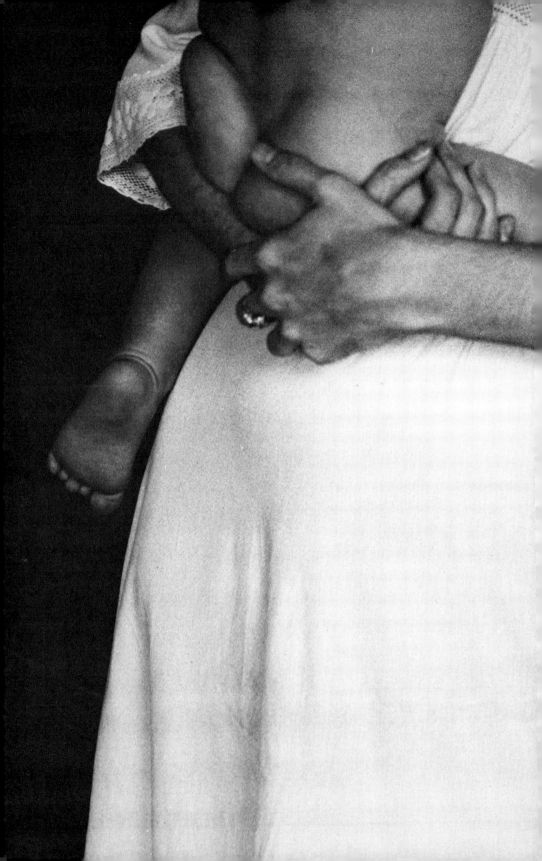

4

Prenatal Environment

Student objectives

1. Describe the importance of the prenatal environment on the developing organism. Include factors such as maternal nutrition, drug intake, illness, radiation, and the mother's emotional state.

2. Differentiate the three stages of prenatal development: germinal, embryonic, and fetal.

3. Identify critical periods in the prenatal state of development.

Assignments for this unit

1. Read Chapter 1 (pages 33–65) in *A Child's World*.

2. View Program 4, "Prenatal Environment."

3. Read Overview 4 in this book and review the study aids. The overview for this unit is a very short synopsis of the material in the text and television program and will not provide all the information you will need to answer the review questions.

Overview 4

Which comes first, the chicken or the egg? Biology students have an answer to this question: Two chickens come first. Together they produce one fertilized egg.

Human development begins when a male sperm or *spermatozoon* penetrates the wall of a female egg or *ovum*. Each sperm and egg is, in a sense, incomplete until this union occurs. Formed by a special process of cell division called *meiosis*, spermatozoa and ova are cells that contain only 23 chromosomes, half the usual number for a human cell. Each of these chromosomes carries the *genes* that, by the combinations they form with the genes of "matching" chromosomes, will determine the characteristics of the infant-to-be.

By a process of "chromosomal recognition" that is still not completely understood, the sperm's chromosomes pair with corresponding chromosomes in the ovum to form a single cell. This cell, called a *zygote*, has the normal complement of 46 chromosomes. Throughout the nine months of a normal pregnancy, the zygote and the cells it forms will divide and multiply to emerge at last as a full-fledged human infant.

The appearance of the zygote begins the first, or *germinal*, stage of pregnancy. The beginning of pregnancy—the union of a spermatozoon with the ovum—occurs when the ovum is traveling through the mother's fallopian tubes. Approximately two weeks later, at the end of the first stage, the zygote will have become a *blastocyst*, or fluid-filled sphere of about 150 cells, and will float in the mother's uterus. At this stage the cells already exhibit structural differentiation, which will be tied to later development.

The implantation of the blastocyst in the wall of the uterus begins the second, or *embryonic*, stage of pregnancy which lasts from the third through the eighth week. During this time the embryo will become dependent on the mother for nourishment and the removal of wastes through the *placenta*. An organ formed in the lining of the uterus by the union of uterine and embryonic membranes, the placenta is attached to the embryo by the umbilical cord. The infant and mother will remain attached in this way until the baby is born and the umbilical cord cut.

At the end of the embryonic stage, major body systems have emerged and several million cells have been formed. Because the fetus is now recognizably human, the third or *fetal* stage is characterized by some people as the beginning of life. From this stage until birth some 32 weeks later, the fetus will grow from less than an inch in length and an ounce in weight to be an average of 20 inches long and 7 pounds at birth. This increase springs primarily from two different types of growth; the cells will increase both in number and in size. The cells of the developing being do not simply reproduce and grow, however; they are multiplied into many different kinds of tissues and organs. When the infant is born, it will have become a composite of brain, blood, fat, muscle, and bone, a complex set of systems designed to function outside the original womb environment and there to grow and to change.

The individual infant that emerges at birth has already been shaped by many factors and influences. Some of the most basic are those carried by the genes; these are the characteristics the infant has inherited from his or her forebearers. Others relate to the infant's environment before birth.

We are not sure of all the things that can cross the placenta from the mother to the infant—and vice versa—or of the effect that these have on the fetus. We do know that the fetus gains its nutrition from the mother and that if the mother is severely malnourished there will be effects on the baby. Similarly, if the mother's diet is lacking in some specific substances, such as iodine, the infant will be affected. Although actual blood does not cross the placenta, substances besides nutrients can be communicated to the fetus if they are carried in the bloodstream. Nicotine is one of these, and mothers who smoke are apt to have smaller children with greater complications at birth than nonsmokers. Drugs such as thalidomide and tetracycline clearly affect the future infant and some diseases—notably measles or *rubella*, syphilis, tuberculosis, polio, chicken pox, and mumps—can either have detrimental effects or be communicated to the child. The effects of alcohol and narcotics can be detected in the fetus, and babies can be born drug addicts with all the withdrawal symptoms of adults present upon birth. But unless the condition is so severe that the infant does not survive, scientists are not sure of the long-range implications for the child.

Fetuses have, under experimental conditions, responded to noise and to the mother's emotional state, but, again, scientists are not yet sure of the long-range effects of these influences. One researcher has tied maternal stress to cleft palates and harelips, and children born to hyperanxious mothers tend to be irritable and restless.

There are periods during prenatal development when the effect of outside influences on the embryo and fetus are greater than at other times. During the first three months of pregnancy the basic organ systems are being formed and the fetus seems particularly vulnerable to drugs, radiation, maternal stress, and malnutrition. It is also during early pregnancy that spontaneous abortions occur for an estimated three out of every ten conceptions, although it is not clear how many of these are a result of environmental insult rather than hereditary influences.

During the second trimester the fetus becomes responsive to loud noises, probably a reflection of its developmental state. During the ninth month of pregnancy, it receives antibodies to a number of diseases from the mother's blood. These antibodies can cover illnesses such as whooping cough, measles, German measles (rubella), and influenza; and they protect the infant for a period after birth. The lack of such antibodies is a source of vulnerability for premature babies.

The existence of differing responses has led to the interesting and complex concept of *critical periods*, or the times during pregnancy when an embryo or fetus is at peak susceptibility to outside influences. It is possible to link the effects of thalidomide to deformity of the limbs or features that were being formed when the drug was taken by the mother. Rubella is particularly dangerous during the first four months of pregnancy, and the hormone DES may have far-reaching effects if taken during this time. Exactly which periods are critical for which aspects of the fetus' growth, however, is not yet known.

Study aids

Review Questions

1. An ovum, which has an X sex chromosome, is about to be fertilized by one of four spermatozoa. Three of the spermatozoa have Y sex chromosomes and the fourth has an X sex chromosome.
    a. Is the child more likely to be a girl or a boy?
    b. Why?

2. How are zygote, blastocyst, embryo, and fetus related?

3. How would you distinguish them?

4. If we represent the nine months of pregnancy as a continuum:
    Conception _____ Birth
    Where on this continuum would the following lie?
    a. blastocyst
    b. placenta
    c. the Moro (startle) reflex
    d. embryo
    e. spontaneous abortion
    f. second trimester
    g. germinal stage

5. An experimenter has found that a mouse given X-ray radiation 7 or 8 days after conception was likely to have a pup with brain hernia, whereas if the radiation was given 9.5 days after conception, the pup might have *spina bifida*.

a. Why might there have been this difference in the effect of the radiation?

b. What relationship might this experiment have to human development? What problems could there be in drawing conclusions from this relationship?

6. What are the possible implications for an infant of a father who takes LSD? How do these differ from a fetus whose mother is a heroin addict? If malnutrition is a state that often accompanies narcotics addiction, would this add possible complications for the development of the child?

Questions Pondered by Psychologists and Scientists

1. By what process can a single cell become a human being?

2. Which environmental influences affect the development of the fetus? When? How? Why?

Of Policy Issues and Public Interest

The issue currently of greatest popular concern relating to prenatal development reaches far beyond psychologists and scientists. It concerns abortion and the morality of taking human life. The question usually asked—When does life begin?—is misleading. It does not define life, and it assumes that any kind of life is worth preserving. According to some persons, it also presents an issue of life or nonlife in isolated, and therefore, too-simple terms. What of others'—for instance, the mothers'—lives? Is life consciousness? Is life a matter of organic existence or is there some other parameter?

Until more information is available, decisions regarding these questions must remain largely a matter of individual choice within the confines of legislation, public opinion, and social and religious sanction.

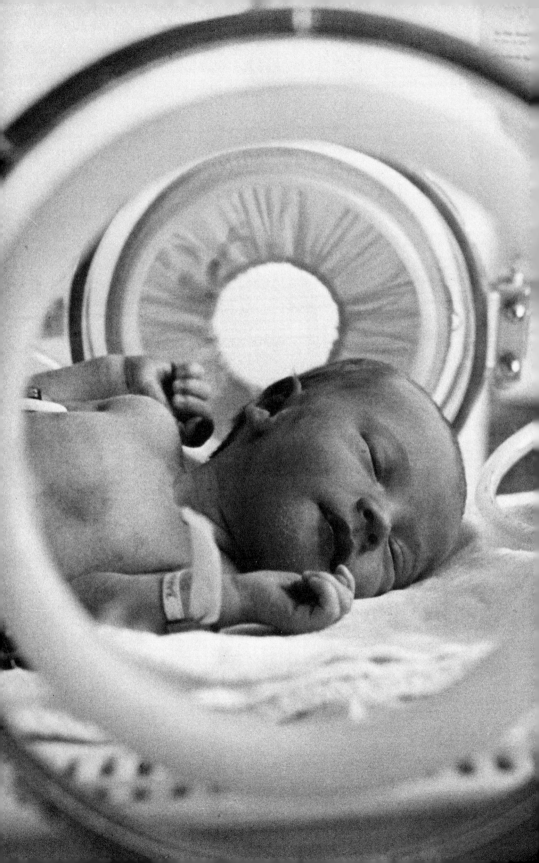

5

The Newborn

Student objectives

1. Identify indicators of normality and describe the purpose of the Apgar scale.

2. Cite ways in which infants demonstrate observable individual differences in behavior from birth (for example, sucking, sleeping, crying, activity, emotion, and temperament).

3. Describe some possible immediate and long-range effects of the birth process and circumstances surrounding birth on the organism.

4. Describe some typical characteristics of prematurity.

5. Describe some possible correlates of prematurity to maternal attachment.

6. Identify reflexes and sensory capacities present in the human child at birth and suggest the relative importance of each.

Assignments for this unit

1. Read Chapter 3 (pages 101–139) in *A Child's World*.

2. View Program 5, "The Newborn."

3.   Read Overview 5 in this book and review the study aids. Although the overview is short, it contains information that is not in the text.

4.   Read Neurological Development in the Newborn (optional).

Overview 5

This unit focuses on the process of birth and the first two weeks of life. This is the period of transition from the mother's womb to the outside world. It is a period of relatively high physical risk for the infant under the best of circumstances. Throughout most of history and in much of the world today it has been considered an extremely critical time for both mother and child. Many people would think the American custom of baby showers—giving gifts before safe passage through this event—a temptation to the forces of evil.

The normal infant begins the process of birth when the mother enters labor. Successive contractions of the uterus force the child from the womb through the cervix into the birth canal and, finally, out of the mother's body into the hands of those who are attending her. The baby is followed by the umbilical cord and the placenta which, together, are called the *afterbirth*. By this time the umbilical cord has been cut and tied at the child's navel and the tiny boy or girl is a separate, if still dependent, little human being.

The newborn infant in a modern hospital is then "checked in." Depending on hospital policy, it may be given eyedrops and a blood test, have its footprint taken, and have an ID bracelet put on its wrist. It is also checked, through a number of indicators, for any problem. In many cases the *Apgar scale* is used. According to this simple but effective scale, the normal infant has a heartbeat of over one hundred, a healthy pinkish skin tone, good muscle tone, strong breathing—often indicated by loud yells—and healthy reflex action in terms of crying, sneezing, or coughing. If there is a deficiency in any or all of these areas, remedial action may have to be taken.

The differences in infants go beyond those indicated by the Apgar scale. The combination of heredity and of variations in the prenatal environment result in distinctive little individuals. Some are placid, some are active. Little Susie's thumb went into her mouth the minute she was put down by the nurse; Donny yelled until he was exhausted. Individual differences such as these will be explored in greater depth in Unit 10, but it is important to note that they begin to determine a baby's interaction with the world around it. The patterns of interaction will, in turn, reinforce the differences the infant brings into its life, further tangling the complex mix of heredity and environment.

Reflex and Sensory Capability

Helpless as it may seem, the newborn infant already possesses a number of capabilities. Responsiveness to environmental events, patterned motor activity, and a number of reflexes are evident at birth. Among other things, the infant can open and close its eyes in response to stimuli, move its eyes in coordination to follow a moving object, and demonstrate convergent eye movements when fixing on a near or distant object. He or she can suck, smile, yawn, grimace, cry, sneeze, swallow and vomit, and move arms and legs. There are other reflexes that are found only in infants. These disappear after a certain period. A list of them appear in the text on pages 108 and 109.

The abilities and reflexes of infants may be divided into four general categories:

1. Sucking, crying, sneezing, and swallowing are among those that are extremely important for infant survival. In this stage they are called reflexes because they are activated by specific stimuli. Later in life the infant will be able to exert varying degrees of control over them.

2. Tracking moving objects with coordinated eye-and-head movements, turning the head toward a source of sound, or turning the head to respond to a touch on the cheek can be

regarded as the primitive beginnings of conceptual and exploratory skills.

3. The swimming, walking, and placing reflexes are among responses that seem to serve little purpose and later drop out of the infant's behavior. They may, perhaps, have filled some function in humanity's distant past. Some workers, for instance, believe that the grasp (Darwinian) reflex was a clinging response, enabling the child to hold onto the mother. The Moro (startle) reflex may be another type of response to danger. The function, if any, now served by these reflexes is not clear.

4. The last category of movements includes activity that is general and that expresses discomfort or arousal. These movement patterns are not adaptive in the sense that the others might be, but they do seem to serve an expressive function in communicating various emotional states to the mother and, therefore, may be important for later personality and emotional development.

Prematurity

It is obvious from the descriptions of infants' capabilities that humans come into the world relatively unprepared for an independent life. Babies who are premature are even younger than the average neonate. Therefore, they are even more dependent on the proper kinds of support from their new environment.

Many premature babies have not had a chance to acquire the antibodies against disease that normal babies receive in the ninth month of pregnancy. Their body weight is lower and fatty deposits smaller, decreasing their ability to regulate temperature. There may be difficulty in breathing. If an infant is very premature, the sucking reflex may not be present, necessitating alternative means of feeding. This is most commonly done through a nasal tube; fortunately, premature babies can swallow.

These factors also mean that many premature babies must be placed in incubators where the temperature can be regulated and oxygen-rich air provided. Unless the baby is extremely premature, chances for survival and a normal life are good. Physical development occurs as it would in full-term infants and the various reflexes appear in the proper sequence. By the time the premature baby would have been a month old if it had been full term, it is generally indistinguishable from a month-old, full-term baby. Equivalent development has taken place if one counts age from conception rather than birth. One might expect the premature baby to be more advanced because it has been in the world longer. On the other hand, the full-term infant is subjected to the trauma of birth when it is better prepared. The processes underlying maturation are complex and little understood, but one chief characteristic seems to be that the many separate systems of growth compensate for and regulate each other. Growth is usually achieved within certain narrow limits of "normality" even in the face of severe environmental hazards.

There is, however, one aspect of prematurity that may have far-reaching effects on the child's development. This has to do with *maternal attachment* or the feeling of the mother for the infant. A few experiments indicate that some parent-child relationships may not be as effective with premature children as with full-term babies. There are a number of possibilities that explain the outcomes of the experiments in this area. Retaining the premature child in the hospital may disrupt a normal process of attachment formation; mothers may "distance" themselves from a premature baby, fearing to love it because of the possibility of loss. The bond between mother and child is more complex and less automatic than has been believed in the past, and more work needs to be done in order to explore the nature of maternal response.

Other problems of the neonate associated with birth can be divided into two categories. The first contains those conditions that developed during the prenatal period and must be corrected at birth in order to ensure a healthy child and mother. These, many of which are detailed in the television program, include heart problems, some genetic diseases, and conditions, such as syphilis, contracted during pregnancy.

The other category of problems associated with birth are those caused by the birth process itself. An about-to-be-born baby may have become entangled with the umbilical cord or refuse to position itself for birth in the womb. Long and difficult labor may result in severe oxygen deprivation, or the use of forceps can result in injury to the child.

Some of these problems can be corrected by variations in the hospital routine such as a Caesarian operation (so-called because Julius Caesar was delivered in this manner) or turning the mother on her side to change the position of the umbilical cord. Others make life more difficult for the newborn and, in some cases, result in permanent injury. Severe birth trauma is easily identifiable and often carries long-range implications; it has been related to reading retardation and impairment in abstract verbal ability, perceptual skills, and social competence. Many other children who have *perinatal* difficulty, or less severe birth trauma, however, show little evidence of long-range effects, although their test scores are different than those of "normal" infants in the earliest days of life.

Study aids

Review Questions

1. What purpose might the *fontanels* serve during labor and delivery?

2. How do the rooting and the grasping reflexes differ in importance to the newborn infant?

3. Pamela had her baby five weeks before term, and it was kept in an incubator in the hospital. Three weeks later she asks:

a. Will her child always be behind full-term babies?
b. Why doesn't she feel the same toward this child as she did toward his elder sister?

What possible answers might the material in this unit provide to these questions?

4. What does a rating of 10 on the Apgar scale indicate about a neonate?

5. Does a newborn baby have depth perception? Describe an experiment which deals with this question.

6. Why is there some uncertainty about the long-range effects of birth trauma?

7. How "old" is a baby born one month before term?

8. If a doctor, who, through some unlikely circumstance, did not know the child was premature, were attending the delivery:
a. What physical indications might there be of prematurity?
b. What steps might be taken to protect the infant and why?

9. What is "natural" childbirth? If you know anyone who has had a baby using this method, ask for her opinion of its advantages and disadvantages.

Questions Pondered by Psychologists and Philosophers

1. What are the origins of maternal love? Do they have to do with hormonal factors? Are they a result of instinct?

2. Is there a quality in the relationship of mother and child that is not replicated in the relationship of another caretaker to the child? What about adopted children?

3. Is there a relationship between the "maternal instinct" of animals and that of humans?

Of Policy Issues and Public Interest

A number of American hospitals have commonly separated newborn infants from their mothers just after birth and placed them in a nursery. Babies were brought to the mother for feeding and then returned. Hospitals felt that mothers and children both needed rest and that they could provide a safer environment for the newborn child in the nursery. There was less chance of infection since few people had access to the children; the temperature and physical environment could be controlled more easily. Some people, however, prefer "rooming in" or having the infant placed with the mother almost immediately. They feel that the health factors may not be as great a concern as they were thought to be in the past. The proximity of mother and child aids breast feeding, and some recent experiments have indicated that infants (and babies) who are cuddled and given attention from birth develop more satisfactorily than those who are deprived of such attention. Mothers differ in which arrangements they prefer. Hospital administrators have noted that the woman who is having her first child is most often the one who wants a "rooming in" arrangement. Women with more than one child, on the other hand, seem to welcome two or three days' rest.

Neurological Development in the Newborn

Neurological development is one of the least understood of the areas of physical development. Many of the changes in the nervous system are too subtle to be observed. Even when changes in structures of cells are noticed, their functional significance often remains mysterious.

A few gross features of brain growth were all that was known to early workers in this field. At birth, the brain averages about 25% of its later weight in the young adult. In contrast, body weight at birth is only 5% of the weight at adulthood; from this we can infer that the infant has a greater proportion of brain relative to body than the adult. This information is interesting, but research has not shown a direct relationship between the size or weight of the brain and a person's intelligence.

A number of sophisticated techniques for studying the brain now provide us with considerably more information than this. Still, our knowledge almost literally only scratches the surface of the cortex. Most authorities today agree that it is probably in the richness of interconnections in the cerebral center that the human brain is unique, although little is known about what goes on in the brain that is related to what is called thinking and ideation. The sequence of development of reflexes and the timing of their dropping out of the infant's behavior provide some indication of neurological development. Abnormal time sequences may be a sign of abnormal neurological development.

Generally most researchers agree that certain regions of the brain serve specific functions. The *cortex* seems to be the structure most highly developed in humans and most closely linked to perception, thought, and language. It consists of a crinkled mantle of millions of cells some two millimeters thick, which is folded in on itself in order to fit into the skull. If spread out flat, the human cortex would cover a medium-sized desk.

In the newborn human cortex, the *neurons* (cells of the nervous system) are only thinly connected with each other. The number of connections that a neuron makes with other cells in the cortex is an important aspect of growth; though it is hard to discover exactly which links have been made and what each means, the interconnecting network becomes thicker as maturation occurs. This change in the cortex is the most obvious during the earliest years of human development.

Not all areas of the cortex develop these connections at the same time. The first regions to experience this kind of growth are those concerned with the visual and auditory systems. The tremendous increase in an infant's responsiveness to sights and sounds is obvious and may be a reflection of this cortical development. At the same time the motor cortex, the region that serves to initiate and organize movements of all kinds, also shows an increase in the complexity of interconnections.

At about two years of age, the so-called "association areas" begin to develop thicker branching. By age four these areas have caught up with the sensory and motor areas. It is assumed, though not yet clearly demonstrated, that these regions link or associate visual, auditory, and other sensory inputs. These are the linkages that have traditionally been considered the basis for all learning, but we can only watch this formation and experiment to discover the abilities that may emerge. The gap between the knowledge of physiologists, who study the brain as a visible piece of matter, and the psychologists, who study the behavior of the whole organism, is still unbridged.

6

The Growing Infant

Student objectives

1. Describe the characteristics of physical and motor development in the child in this period.

2. Using walking as an example, describe differences in the rate of development in walking as a function of cross-cultural differences in child rearing.

Assignments for this unit

1. Read Chapter 4 (pages 141–167) in *A Child's World*.

2. Read Overview 6 and the section on Children Across Cultures in this book. Note that the overview treats some subjects that are not in the text and that the last section updates some of the text information.

3. View Program 6, "The Growing Infant."

4. Review the study aids.

Overview 6

A two-and-a-half-year-old child bending over his new-born sister illustrates some of the tremendous changes that occur in physical development after a child is born. More than change in size and weight alone, this development has brought together the various systems of his body into a functioning whole.

Growth in Body and Capability

Physical appearance indicates many of these changes. The newborn infant's skin is red and crinkly from being surrounded by fluid in the womb. The head is large compared to the rest of the body. Movement is restricted; the infant cannot yet hold up her head, let alone sit or walk.

As the newborn girl follows her brother through the first years of life, her downy baby hair will drop out and be replaced by coarser hair, closer to the color she will have as an adult. Gradually her soft skull and skeletal bones will harden or *calcify*, giving her bone structure the strength she will need to carry her body erect. The skeleton will also grow faster than her skull. As it does, her proportions will change, as illustrated by Figure 4-1 (page 142) in the text, and her center of gravity will shift downward. As a toddler, however, it will still be at about her chest—unlike an adult's whose center of gravity is at the pelvis—causing her to walk with her legs set wide apart for balance.

As we have seen in Unit 3, she did not arrive totally unequipped for life. During her first few weeks many of the reflexes with which she was born will disappear. Some claim that these are simply "forgotten" by this infant because she has no opportunity to practice. A more widely accepted explanation is that growth of the nervous system is responsible. These early responses—"walking," "swimming," "grasping"—are called *reflexes* because the voluntary aspect is absent; they seem to be carried out automatically when the proper stimuli are present. As a child matures and the higher nervous centers come into play, the lower mechanisms, including many reflexes, may be shut off.

Over the next years her development will follow patterns common to all normal human infants. It will be *cephalocaudal*—the areas nearest the head will develop first—then *proximodistal*—regions near the midline will develop more rapidly than those away from the midline. She will move through certain stages of growth that show little variation in sequence. Growth will occur in a coordinated fashion, in many different systems.

Maturation and Readiness

An important fact in physical growth is that the nervous system, which mediates behavior, develops together with the skeletal muscles and bones. A normal infant does not wait for months before toddling across the room simply because his bones are too weak to support his weight. He cannot walk for neurological reasons as well. The *cerebellum* and motor cortex of the brain, both concerned with the control of body posture and movement, are still immature at birth. They grow rapidly in size and complexity and, about the time an infant's muscles and bones are ready, have developed to a point where they can coordinate crawling and walking. The walking motor pattern generally appears about 12 to 15 months after birth.

A further complication in walking is that this "simple" act requires the coordination of many muscle systems, which do not all mature at the same time. Several subskills are involved in walking. The head must be held upright, the arms must be capable of supporting the body through holding on to things during the initial period, and the legs must be coordinated enough to move alternately while supporting the body weight.

Another important type of motor control that develops over the course of several months involves skill in looking. The eyes must move together. Looking also requires that the sensitive center of the retina is directed at the object of attention. New born infants possess good eye coordination but do not use their eyes well in picking up visual information. In a sense they do not know where to look or what part of the optic array to sample.

It may be that even aspects of perceptual and cognitive development can be related to the maturational processes of the nervous system. Since researchers cannot systematically control the environment and upbringing of infants, definitive evidence is difficult to obtain, but we do know that there is a close link between language acquisition and motor development. If an infant is greatly retarded in crawling or walking, there is a good chance that language is underdeveloped as well.

In this complex coordination of muscles and muscle systems via the central nervous system, the appearance of sequences of motor skills suggests that maturation is the key process rather than learning. *Maturation* implies that inner changes are taking place more or less independently of practice. If learning were the crucial process in motor development, skills should appear through practice, but this does not happen. Although we do not yet understand the nature of the critical influences on motor development, we know that practice is not one of them at this stage.

What then could be the basis for this type of maturation? The cells of the infant and very young child provide one clue. There are fewer cells than in the adult, but though our newborn baby's cells will multiply four times before she is entirely grown, this factor does not seem to be the critical one. A much greater amount of growth and perhaps a more crucial type will occur in the interconnections among cells in the brain. These increase greatly with age and, as outlined in the optional reading for Unit 5, appear to be connected to mental development.

Another change that will occur within a few months after birth is growth in the ability to regulate internal states and processes of the body such as temperature and digestion. These are closely connected, and changes in one influence the others unless the change is regulated. Temperature, for instance, affects the speed of biochemical reactions; yet it is only after the first month of life that the sweat glands begin to operate, to provide a cooling system in warm environments.

The sleeping pattern is irregular at first but becomes predictable with age. Feeding and *evacuation* also become regularized. Even breathing, fast and irregular after birth, slows and becomes regular. Proper control of the internal environment is essential for the later maturation that gives a child the ability and desire to seek new experiences in her external environment.

Study aids

Review Questions

1. What factors connected with neurological and physical development prevent an infant of two months from lifting a heavy object?

2. Differences in height and weight between children from poverty-stricken homes and children from economically advantaged homes are evident as early as the first year and remain consistent throughout life. Cite some possible causes of such differences.

3. Cite evidence that motor development cannot be accelerated. Is there any evidence that some activities *can* be accelerated? Explain your answer.

4. How do walking and grasping relate to
   a. *Cephalocaudal* development
   b. *Proximodistal* development

5. Find an appropriate place for each of the following events on this continuum.
   Birth _____ One year
   a. Crawling
   b. Rolling over

c. Grasping of objects using thumb (prehensile)
d. Lifting head

6. How do the proportions of head and body change during the first two years after birth? What implication does this have for walking?

Questions Pondered by Psychologists and Neurologists

Is there a built-in mechanism which regulates the maturation process and compensates for slow development at one point by speeding up at another time? What kind of mechanism could be involved?

Of Policy Matters and Public Interest

Many parents take their children to baby exercise classes. Advocates of the use of exercises at a very early age say that it aids growth by using the movements the baby would make naturally. They add that it helps establish warm ties with parents through the interaction the exercises require. Others hold that children will develop naturally and that, indeed, since physical development is linked to maturation, that babies are not ready for some of the exercises they are required to do. When this happens, they say, it leads to frustration and possible impairment of personality development.

Children across Cultures—Walking

The age at which an infant begins to walk is determined by many factors. One of the most important is genetic—not simply whether a child is hereditarily predisposed to walk early or late but those genetic factors that influence growth, size and weight, and rate of maturation. Linked with these are sex—girls tend to walk earlier than boys—and nutrition. Also important are cultural factors—whether the child is encouraged to roam about or bound with swaddling clothes, the amount of contact with the mother and other individuals, and related social attitudes.

Whatever the differences, normal infants eventually learn to walk. By the preschool stage, it is almost impossible to tell who was an "early" and who was a "late" walker. If this is the case, why attach so much importance to the age at which walking started?

Part of the answer, of course, lies in psychologists' and physiologists' desire to increase their understanding of the maturational process. Whether the age at which a child walks is primarily genetically or culturally determined, however, has become a social issue as well.

In the 1950s French psychologist M. Geber published a number of articles concerning African children. Geber's studies in Uganda showed that Ugandan infants' motor development was considerably advanced compared to norms based on white children. Geber concluded that the difference was probably genetic in origin.

Based largely on his studies, a number of scientists then argued that this was another indication that blacks were mentally inferior to whites. They pointed to the fact that, in general, the higher the species is on the evolutionary ladder, the longer the period needed for psychomotor development. They reasoned that if blacks matured more quickly than whites, blacks were a lower order of mental development.

This hypothesis, of course, could not be tested directly. Furthermore, a number of subsequent studies have indicated that the acceleration of psychomotor development among certain groups of infants—largely those in preindustrial societies—seems to be culturally based rather than genetic in origin. A recent survey of such studies concluded that the infants who showed accelerated psychomotor development shared common experiences during the first year.*

*Emmy E. Werner, "Infants Around the World: Cross-Cultural Studies of Psychomotor Development from Birth to Two Years," *Journal of Cross-Cultural Psychology*, vol. 3, no. 2 (1972).

They were members of an extended family system with many caretakers; they were breast fed on demand; they were constantly in contact with an adult caretaker; they did not wear restrictive clothing; and they ate, slept, and were toileted on demand rather than according to a schedule. The studies of infants of the same ethnic group who were raised in a "modern" or "Westernized" environment did not show such a degree of psychomotor acceleration.

This indicates that differences in rates of psychomotor development between certain black groups and Western whites reflect environmental differences in child-rearing rather than genetic differences. The postulated link between intelligence and psychomotor development is shown to be unlikely by this evidence. But in any case, any such link would have to be demonstrated by a variety of studies, not just a single correlational study which alone cannot prove an underlying causal relationship.

7

The Learning Infant

Student objectives

1. Define the concept of learning and the various conditions under which learning takes place, using the following research-theory orientations: building mental structures, classical and operant conditioning, and imitative learning.

2. Discuss how the infant is an active, exploring, information-seeking organism from birth, and illustrate how these attributes can be encouraged or discouraged.

Assignments for this unit

1. Read pages 169 to 188 in *A Child's World*.

2. View Program 7, "The Learning Infant."

3. Read Overview 7 and Object Permanence in this book.

4. Review the study aids.

This unit begins the first of a set dealing with mental or *cognitive development*. In it we will introduce theoretical perspectives, which will be continued in Units 13, 19, 20, and 27 and which will deal with mental development in the preschool child, in the middle years, and in adolescence. This unit also continues a demonstration of the fact that it is not easy to divide the subject of child development into compartments. We have seen the interaction of heredity and environment carried through the prenatal and neonatal periods; it will continue to appear throughout this course. Similarly, you will find that some of the theories presented in this chapter appear in other areas, particularly in the units on personality development and language learning.

The question of learning has intrigued philosophers and psychologists for centuries. What is learning? How does it occur? Although, logically, it would seem that the first question must be answered before the second could be addressed, the major portion of this unit deals with theories about the process of learning. Each springs from an approach to the concept of learning, but none of these approaches has provided psychologists with a completely satisfactory definition of the term. We will, for the purposes of this course, use the definition provided by the text: *Learning* is "the establishment of new relationships or the strengthening of weak relationships between two events, actions, or things." Students should be aware, however, that this is not a definitive characterization.

British philosopher John Locke (1632–1704) was one of the first people to study—or write about—learning in infancy. According to Locke, a child was born a blank slate or *tabula rasa* with neither knowledge nor distinct perceptions. In Locke's view learning occurred as the environment scribbled on the child's mind. He would receive sensations from the people, events, and things around him and by virtue of the fact that certain sensations recurred regularly, would begin to combine them. Reliable clumps of sensations would come to represent stable objects and features of his world. Eventually, as they grew more complex, and as the child grew older, these clumps would form ideas.

Because of the emphasis on experience, Locke's view was called *empirical*. We now know that this perception of a baby's world is not correct. The neonate (Unit 5) enters the world with certain reflexes, and recent experiments indicate that his or her perceptions of the world are much clearer than Locke's theory would allow. T.G.R. Bower, a British psychologist, "taught" six- to eight-week-old infants to turn their heads when he placed a cube in front of their eyes by having their mothers pop into sight and say "peekaboo." Once this head-turning was established, Bower substituted a different size cube. The infants turned their heads less readily. Then he used the original cube but placed it at a difference distance, and the infants turned their heads as frequently as they did before.

Bower concluded that infants could distinguish the *optical* size of the cube from the actual size. They could tell when the cube appeared to be smaller or larger because it was a different distance from them, from the cases where the cube actually was smaller or larger. Since it seems impossible for infants to learn the properties of space in such a short period after birth, this experiment contradicts the classical assumption about their minds being blank slates.

Behaviorist Theory of Mental Development

Locke's general approach to learning, however, is continued by the *behaviorist* school of thought. Like Locke, behaviorists believe that the infant, though obviously not a tabula rasa, learns primarily from its environment. Another term for this approach is S–R, or stimulus–response, theory since these psychologists assume learning consists of the linking together of stimuli and the responses to these stimuli.

Stimulus–response theory originated with a Russian psychologist during the early twentieth century. Ivan Pavlov, who received the Nobel Prize for his work on digestion, happened to notice that his dogs would salivate and produce stomach contractions when the experimenter entered the room, rather than when the dog could see the food he carried. Instead of dismissing this behavior as an experimental nuisance, Pavlov realized its great significance for psychology and dropped his work on digestion to study it.

In the cortex of the dog, according to Pavlov, there is a center that receives the signal corresponding to the stimulus. If this center is aroused, it leads to the natural response of salivation—the hungry dieter's mouth waters at the sight of a warm, crumbly brownie fresh from the oven. This is a natural or *unconditioned response*. If the food stimulus, however, is always presented with another, irrelevant, stimulus such as a bell, the excitation center that responds to the sound is gradually, by constant repetition, linked to the center that causes salivation. Once this linkage is made, salivation occurs when the bell sounds whether or not the food is present. This behavior is known as a *conditioned response*. Most animals, including humans, can be trained to give conditioned responses.

Pavlovian conditioning is often called *classical conditioning*. The response is involuntary, but it is learned, unlike a response such as salivation at the sight of food. Another type of conditioning is associated with an American psychologist, B.F. Skinner. Skinnerian or *operant conditioning* occurs when the organism—animal or human—acts on the environment and its action results in a reward or punishment. A well-known example of operant conditioning involves a rat and a lever. If the rat is rewarded with a pellet of food when it pushes the lever, it will push it often. If, on the other hand, the rat receives a shock when the lever is pushed, it will push it less frequently. An example of operant conditioning that is closer to the subject of cognitive development in infancy is the baby who learns that someone will come to take care of her if she cries. In both cases the subject acts on the environment in a particular way to obtain a desired result.

Imitative Learning

Imitative learning is another example of learning that is directly related to interaction with the environment. Strictly speaking, imitative learning is less a fully developed theory of learning than a term to designate certain types of behavior. The infant will imitate movement and human sound, and this imitation can be reinforced through interaction with adults. Imitation can be seen as a means of increasing a child's repertoire of action.

Piaget and Mental Development

Jean Piaget, a Swiss psychologist, has developed one of the most fully articulated theories of learning in child development. According to the Piagetian theory of building mental structures, innate capabilities interact with the environment to lead a growing child through several stages of cognitive development. Infants, to about two years of age, are in the *sensorimotor* stage, described in detail in the text. From two to seven years of age, most children are in the stage of *preoperational* thought. At about seven they enter a period of *concrete operations* which lasts until the attainment of *formal operations* in the early teens.

These stages are sequential and none can be skipped in development, but they also overlap to a degree. Elements of earlier stages of thought may remain throughout life.

In Piagetian thought, intelligence is a process of adaptation to the world. Adaptation occurs through two related processes, *assimilation* and *accommodation*. Assimilation, as in the incorporation of food by the stomach, refers to the taking in of information from the environment. This input is "absorbed" by an internal (mental) structure called a *schema*. Accommodation, exemplified by changes of the organism, such as chewing, stomach movements, and excretions of juices, in order to accept and digest food refers to the *changes* in the mental schema caused by assimilation of new information. This circular process ensures adaptation to the realities of the world. The principles of assimilation and accommodation, Piaget believes, describe the functioning of biological systems in general, whereas the schemata represent internal structures or the mental organization of the specific organism, in this case, the child.

Long before the child can talk, it interacts with people and objects in its environment. These interactions and the infant's changing responses to stimuli indicate that the infant learns to anticipate events, adapt actions to different events, and form "concepts" of a nonverbal sort. According to Piaget, this period—the *sensorimotor* period—lays the foundation for both language and thought.

Infant behavior during this period progresses through several stages. At stage 1 the infant only relates its movements to its alimentary needs and does not properly distinguish its body from surrounding surfaces. In stage 2 *primary circular reactions* appear. These are repetitive limb and body actions that the baby makes in relation to itself. According to Piaget, the baby is still not relating to the environment.

With stage 3 the maturational process allows greater motor activity. The infant can reach, crawl, and manipulate objects around it. Actions of this period are characterized as *secondary circular reactions* by Piaget. They differ from the preceding stage in that the infant's actions are directed outward to objects rather than to its own body.

Often the infant will discover that some action will produce a change in its environment. This—the connection between the action and the change—becomes a point of interest. Acting on the world is a new experience for the infant who has been the subject of action to this point. Objects then acquire meaning according to their action–relation to the infant—the object is assimilated to a sensorimotor schema. With familiarity the full action sequence elicited by the object is abbreviated and finally becomes internalized. The schema can now be thought about through the use of a symbol. The use of this symbol in thought can, in turn, activate the sensorimotor schema.

In the later sensorimotor stages the infant engages in more elaborate behavior. The schemata are coordinated so that several actions are organized into a sequence adapted to act on the world.

This goal-directed behavior results in *tertiary circular reactions*. It is a behavior that is intentional, flexible, and novel and aimed at discovering the environment in a deliberate fashion. The children now exhibit interest in objects and events for their own sake, independently of themselves, and are ready to move into the stage of preoperational thought.

In all of these theories the infant must reach out and explore the world in order to learn. Experiments have demonstrated that even very young infants enjoy novel experiences, although an event or noise that is frightening can inhibit interest and learning. The experiments outlined in the text to illustrate operant conditioning in infants demonstrate not only that the babies can learn but that they want variation and stimulation in their environment.

Various skills are developed through infants' information seeking. Reaching and grasping leads to hand–eye coordination which, in turn, will be important for learning to read in the early school years. Babbling is, as outlined in the next unit, a precursor of speech. Lack of stimulation or an environment that does not provide the infant enough security in which to explore can result in learning inhibition.

Study aids

Review Questions

1. How does the experiment with Little Albert relate to the infant as "an active, exploring, information-seeking organism"? (The experiment is on page 184 of the text.)

2. Colin could not find a toy which had been hidden behind another toy while he watched. Does this tell us anything about his probable age?

3. How and when, in Piaget's terms, would Colin come to learn that the toy is under the cloth?

4. Which of the following is an illustration of *classical conditioning?* Of *operant conditioning?*

a. Patting a dog because it wags its tail

b. Fear of a growling dog because of an earlier dog bite

Can you think of other examples of classical and operant conditioning?

5. What is a *secondary circular reaction?* How does it relate to cognitive development in infancy?

6. Describe two experiments which indicate that the infant enjoys novelty and stimulation.

Questions Pondered by Psychologists

1. How does an infant learn to stick out its tongue in imitation of an adult? This fascinating question involves one of the most intriguing psychological problems: How does the infant connect the visual sight of the parent's tongue with its own tongue which it cannot see?

2. Why do humans attain adulthood so slowly relative to their lifespan? What implications does this fact have for the nature–nurture debate? What biological advantage is there in this prolonged childhood?

Of Policy Matters and Public Interest

Two of the theoretical approaches to learning introduced in this chapter—the behaviorist and the Piagetian—are illustrative of differences in parents' and schools' approach to teaching children. If learning is a result of conditioning, it may be considered wise to begin "teaching" infants and children as early as possible. If, on the other hand, learning is a result of a built-in maturational process that occurs in a fixed sequence, an attempt to "teach" before a child is ready would not only be useless but might also be harmful. In fact, many of the teachers who work with children find themselves using both approaches.

The Development of Object Permanence

Until an infant is about two months old, "out of sight" is truly "out of mind." Objects that the child cannot see do not exist. The process of learning that objects do continue to occupy space and time whether or not they can be seen is a long one that raises basic questions concerning the nature of perception and learning.

After about two months of age an infant will follow an object with its eyes and head and will continue to "track" it even after it has gone out of sight. A five-month-old baby will probably retrieve a toy that is partly visible under a cloth, but if the toy is completely hidden, no attempt to retrieve it is made *even if the infant saw Mama put it there*.

An infant who is six months old or older will reach out and pull the cloth off the toy. He knows that the toy exists even if he can't see it, but he is still easily fooled. If the toy is regularly hidden under one cloth and then, in full view, hidden under another cloth in a different location, the infant will reach for it in the *old* location. This "place error" disappears by about ten or twelve months of age.

Full object permanence has still not been achieved. If the cloth and the toy hidden under it are switched with another, flat, cloth, the infant will again make a place error although, to an adult, the presence of the toy under the old cloth is obvious. A child of this age is also disturbed if the experimenter holds the toy in her hand, puts the hand under a cloth, and draws it out empty. Instead of looking under the cloth, the child will look at the experimenter's hand. Only at about 18 months does the object concept appear to stabilize and the child begin to exhibit adult behavior patterns.

Piaget relates these patterns of behavior to the stages of sensorimotor development. He suggests in his description of them that the child "acquires" the concept of object permanence as part of the natural maturational process. An alternative view is that the child learns or perceives the permanent nature of these objects through seeing them from a variety of angles and in a variety of circumstances. A *concept* of permanence is not required.

8

The Beginnings of Language

Student objectives

1. Describe some of the theories of language acquisition.

2. Describe the stages in the early acquisition of language.

3. Relate language acquisition to the child's conceptual development.

Assignments for this unit

1. Read pages 189 to 196 in *A Child's World*.

2. View Program 8, "The Beginnings of Language."

3. Read Overview 8 and The Relation of Language to Conceptual Development in this book. Objective 2 is well covered in the text and television programs, but the information on Objectives 1 and 3 is substantially augmented by the study guide.

4. Review the study aids.

People respond to babies. They smile at them, laugh with them, play with them. Most important, they talk with them from the time a father repeats the first "Dah," to that exciting moment Baby says his or her first real word.

This first interaction, with infant and adult echoing each other's "nonsense" syllables, is usually repeated many times. As it continues, the infant's sounds are more and more similar to his mother's and father's speech. When the infant is around ten to twelve months, an adult may realize that one of the sounds carries a specific, understandable meaning. It is a word. From this time on, the infant's repertoire will grow with amazing speed.

At first, the words appear singly, and the listener has to guess or interpret what the infant means to say. The word "Mommy" may mean "Look, here's Mommy" or "Mommy, I'm wet and uncomfortable" or "Mommy, pick up my rattle." Two-word combinations appear during the second year of life. These may also have a variety of meanings, depending on the context. Three-word "sentences" with regularities in the use and order of words come next, at around two years of age.

By three or four a child will have acquired the capacity to communicate fairly clearly. This represents an incredible expansion of linguistic ability over the course of two short years. No other achievement in development is quite as dramatic or as full of significance for understanding human behavior.

Language and Communication

Fully developed language is, as far as we know, unique to human beings. Animals communicate, sometimes through complex systems, but for one reason or another we do not consider this communication "language." Language goes beyond natural animal expression or communication in a number of ways. By examining the ways animal and human

THE BEGINNINGS OF LANGUAGE

communication differ, we can isolate some of the characteristics of the concept *language*.

A bee who has discovered a patch of flowers communicates this knowledge to its hive by means of an intricate dance. The manner and frequency with which the bee wiggles its body "tells" other bees the distance and direction of the discovery. Why do we consider this *communication* but not *language*?

First, the bees do not "talk" to each other. No questions are asked of the explorer bee either before or after its dance. Communication is one way; it is not conversation.

Second, the dance is *invariant* for all messages. The meaning of a particular move is always the same, and if the bee wishes to communicate a different message, he has to make a different move. There is no grammar in this communication, which severely limits its flexibility. Grammar describes the way in which words are organized in a sentence regardless of what words are used. The actual words do not matter as long as they are the appropriate parts of speech. This allows human speech a great number of combinations of a few words to form many meanings: "The cat bit the dog" and "the dog bit the cat" use the same words, but the meanings of the words is changed when the subject and object are reversed.

Third, human language can refer to situations and events that are not immediate. No animal seems at all capable of conveying a message about something that happened yesterday. It is true that alarm signals—the buzzing of a rattlesnake—indicate that something dangerous is about to happen, but in this example the future is very concrete and immediate.

The fourth and possibly the most compelling difference between humans and other animals is that humans can form abstract concepts or ideas. Animals, no matter how intelligent they seem, do not appear to think abstractly. The reason clearly involves the absence of language, although just what the relationship of language to abstract thought may be is unknown. This particular problem will be discussed at some length in Unit 14.

The Learning of Language

Humans have tried to teach animals to speak; the most successful experiments have been with chimpanzees. A number of systems have been used. Some require the chimps to make sounds; others substitute plastic discs for words or use sign language. The latter methods have been the most successful, perhaps because the chimpanzee's vocal apparatus can only produce a limited number of sounds. One pscyhologist, David Premack, taught a chimp named Sarah with colored tokens. After a period of intensive tutoring, Sarah learned to combine tokens in different sequences to symbolize different objects and the relations among them. In all of the experiments the highest level of language attained by any of the chimps (most of whom are now as well known as the researchers) was approximately that of a two-year-old human child. Even beyond this apparent limit to chimp language development, however, another factor is of great importance in distinguishing chimps' language use from that of humans.

Human infants are not consciously "taught" to speak. They learn, somehow, under all except the most adverse circumstances. Exposure to the sounds of the language is clearly important and the infant seems to learn better with fondling and attention, but no particular training is required. The chimps, on the other hand, did not learn "naturally." Even those who were raised as part of a human family required intensive and continuous teaching before they could use "language."

The natural propensity for language in humans seems to be so great that there have even been cases in which children developed their own language spontaneously. One study concerned deaf siblings born to hearing parents. These children did not have spoken language but developed their own sign language even though they had had no contact with deaf people who used this type of communication. Not only were they not given instruction in sign, they actually taught their parents how to use their special language. This somewhat surprising result supports the view that language is specific to humans but is not restricted to vocal speech.

Theories concerning the acquisition of language begin with observable developments in the use of language in children. These, described briefly at the beginning of this overview and in greater detail in the text and television program, occur in a sequence and with a timing that is relatively stable across cultures and classes. How and why this occurs has been interpreted in a number of different ways by proponents of different theories of development.

The behaviorists' approach to language learning reflects their views of learning in general. Initial learning may occur through *conditioning*. The child associates the sound of the word *cup* with the sight of the cup. He is reinforced by the response people around him give to his use of sounds or language and gradually learns that he can affect his environment through the use of words. By this time, he is learning through *operant conditioning*. The exact course this learning follows has been debated by behaviorists. Some of the different approaches are outlined in the text.

There are some problems in this approach to language learning, however. The deaf children cited above seemed to develop language without this kind of reinforcement from their environment. Furthermore, language learning through a chaining together of stimuli and responses to them does not completely explain the development of grammar. Recent experiments indicate that parents use a simplified form of grammar (or "motherese") with children and that they listen to what children have to say rather than the way a sentence is constructed. Rewards are based on the content of the child's speech rather than the syntax, yet the grammar used by the parents in their own speech is gradually adopted by the child.

At the other end of the heredity–environment spectrum from the behaviorists are many linguists. The theories, variously labeled *preformationism* and *predeterminism*, share the belief that human beings have an innate, biological predisposition for language. A neurological mechanism for language is inherited; the language actually acquired depends on circumstances. The use of rules to govern speech even before "correct" grammar appears—for example, the instances cited in the text in which irregular past tense verbs such as *went* are used in regular forms such as *goed*—adds some support to this view.

Piaget's theory of language development is assigned to the interactionist school because it postulates the interaction of inherited and environmental factors. Piaget's approach, like that of the Behaviorists, reflects his view of cognitive development. Piaget feels that children are born with the capacity and need for language. When a child reaches the appropriate maturational level, she is able to represent objects or people symbolically through the use of words which form, in turn, the basis for further cognitive development in the preoperational stage. This ability and her desire to use it is reinforced by her environment, usually in terms of success in communicating a message. In the early stages of language development this reinforcement probably also occurs through mutual imitation and mutual interaction between child and adults.

Study aids

Review Questions

1. Arrange these stages of language development in the correct sequence:
   a. Babbling
   b. Holophrase
   c. Lallation
   d. Cooing
   e. Multiword sentence
   f. Echolalia

2. How, in the behaviorist approach, is language learning related to environment?

3. Was Paul Revere's code "One [light in the church] if by land, two if by sea," closer to human language or the communication of bees? Why?

4. Zia calls almonds "almonds." They are *not* nuts as far as she is concerned. She calls both walnuts and filberts nuts.
   a. Is she overextending or underextending the word *nut*? Why?
   b. What does this indicate about her conceptual development?

5. What is the LAD and to which theory of language development does it relate?

6. Can a child learn to use language without speaking? Explain your answer.

Questions Pondered by Psychologists and Linguists

1. Does a chimp like Sarah really possess language or is Sarah simply a well-trained animal?

2. Are the restrictions on speech during very early childhood due to the limited capacity of thought or to a limited understanding of how language is put together?

The Relation of Language to Conceptual Development

The early course of language acquisition is related to conceptual or cognitive development (see Unit 7). The development of babbling into words and the subsequent joining of words to form phrases and sentences is a clear indication of the infant's growing ability to impose his order on the environment. The conceptual development that has taken place in the first two years of life—and the development that has yet to occur—is reflected in the two-year-old's handling of words and syntax.

Children tend to *overextend* and *underextend* the meanings of words. This may be a function of limited vocabulary—"doggie" can be used for all animals (overextension); "night" may only occur at bedtime rather than when the sun goes down (underextension). As the child's command of language increases, she will differentiate among different animals ("doggie" is now only used with dogs), but the correct use of class-inclusive words (A Pekingese is a dog, not a cat; dogs and cats are both animals) does not come until sometime later. The specific cognitive processes relating to conceptual development will be discussed in later units. There are also problems in language use which indicate the child is unable to consider all aspects of an object. Young children tend to form concepts and learn words based on functional characteristics—a tricycle is something to ride on, not a vehicle with handlebars and three wheels.

Perhaps even more indicative of the level of conceptual development is the nature of the words and combinations of words used by a two-year-old child. In all languages so far studied the earliest two- and three-word sentences are limited to a rather small set of relations concerning the sensorimotor world—the world the child can see, smell, hear, and touch. The acquisition of plurals, past tenses, and prepositions depends on the development of concepts of time, number, and more sophisticated relations.

9

The Emerging Personality

Student objectives

1. Describe the concept of personality and identify some of its components (for example, aggression, achievement, autonomy, attachment-dependency) and note which components are primarily developing in this period.

2. Compare and contrast various theoretical perspectives for understanding personality development: psychoanalytic, psychosocial, and social learning theory approaches.

3. Describe the development and nature of attachment in this period.

Assignments for this unit

1. Read Chapter 6 (pages 215–255) in *A Child's World*.

2. View Program 9 "The Emerging Personality."

3. Read Overview 9 in this book and review the study aids.

Psychologists have long been concerned with the concept of personality. This, more than any other aspect of an individual, seems to hold the key to why that individual is herself and no other. Personality begins to be apparent soon after an infant is born, but its components are clearly not immutable for the child's personality traits will continue to develop and change into the adult years.

In studying personality the focus is on the person as a psychological entity. The personality may be composed of traits, each of which has varying degrees of influence, but the behavior we study is that of the whole individual. In this study, it is taken for granted that the person shows qualities that are *stable*. If traits and behavior are changed from day to day, it is difficult to characterize a "personality." Furthermore, since a person's character, attitudes, emotional reactions, and moods in social interaction have generally been formed within a specific social setting, it is only from within that setting that his personality can properly be assessed. If assertiveness is valued in the United States and not in Japan, assessing a Japanese personality on the basis of U.S. standards of assertiveness will not produce a true profile.

Personality, then, is the unique blend of traits, behaviors, feelings, and characteristics that characterize an individual. These features are relatively stable over time, and their formation is, among other things, dependent upon the sociocultural background of the individual. Physical qualities are not directly included in the definition of personality though they may be relevant to psychological aspects of the person.

Personality and Infants

For all of the individual differences shown by infants, psychologists hesitate to predict future personality for several reasons. The reactions and expressive capabilities of infants are limited. Their behavior is closely tied to physical state because of their helplessness. Moreover, the fast pace of physical and neurological maturation makes the human one of the most unpredictable of organisms, for the development of these systems affects many things, including personality. Nevertheless, it does seem possible that some infant behavior is an indication of what the adult may be like and studies have supported this possibility although the results have not been definitive. In one, for instance, psychologists observed 25 infants over a two-year period. Fifteen years later another psychologist located 15 of these infants—then teenagers—and developed personality profiles for them. People who did not know any of the subjects then tried to match the infants' and teenagers' profiles. They did have enough success to indicate there was some correlation between infants' and adults' personality traits, but no clear pattern emerged concerning which aspects were most predictable.

Theories of Personality and Social Behavior

Although the events of a person's life affect personality, there is a great deal of discussion as to how this occurs. In this unit we introduce three theoretical approaches to personality that will be repeated throughout the course: *psychoanalytic*, *psychosocial*, and *social learning theory*. All of them focus on the interaction of various forces in the development of the individual, and none of them is totally independent of the others. Each, however, stresses different forces in personality formation and implies a different approach to personality change.

Around the turn of the century Sigmund Freud, a Viennese physician with an interest in hysterical patients, developed a new treatment for mental illness and, from it, elaborated a new theory of psychological functioning. He proposed that the adult psyche consisted of three parts: the *id*, the *ego*, and the *superego*. These emerged, in that order, during development.

The id represents the primitive, pleasure-seeking unconscious, the repository of sexual energy and aggression. An infant's psyche consists only of the id. Correspondingly, an infant's behavior can be described simply in terms of seeking pleasure and avoiding pain or discomfort.

The ego forms as a result of socialization and controls the impulses of the id. It represents the self. As the child develops mentally, he understands reality better and is thereby able to adapt his behavior to outside standards to a greater degree. Pleasure is still sought, but it may be delayed.

Finally, the superego constitutes the conscience. It develops by incorporating parental values and attitudes. The superego maintains ethical standards in behavior by controlling the ego and id. If, for example, the ego, in the service of the id, sees a way to achieve a certain unethical goal, the superego will inhibit action or create guilt feelings if the action is carried out. Because of the superego, individuals come to conform to social conventions. The superego can, of course, incorporate the attitudes of people other than parents, but in the developing child, the parents are the prime authority figure.

Freud characterizes infant personality development by two stages, the *oral* and the *anal*. These are discussed at some length in the text.

Freud's *psychoanalytic* or, as it is called in the text, *psychosexual* theory of development was the basis for Erik Erikson's *psychosocial* approach. Like Freud, Erikson sees conflict as a basis for development, but the conflict is rooted in social interaction rather than sexuality. According to Erikson, eight stages occur during the life span, all of which shape ego development (see Table 6-1, page 221 in the text). Two of these—trust versus mistrust and autonomy versus shame and doubt—occur in infancy. The direction in which each of these is resolved influences later development.

A third approach to the development of personality is neobehaviorist or *social learning*. Infants and children learn from the environment around them. They imitate the significant models in their lives and are reinforced when those models approve of their actions. The nature of the imitation and learning is characterized by the following:

1. When the model is reinforcing, it is more likely to be imitated.

2. When a model is punishing, it is less likely to be imitated.

3. When a model is warm and nurturant to the child, imitation is more likely.

4. When the model has resources, such as food and attention that the child craves, then imitation is more likely.

5. Prestige or status of a model influences imitation.

6. Models are more influential when they are viewed as similar to the one who is being influenced.

7. Age and ethnicity influence imitation learning.

This theory suggests that certain aspects of personality can be learned through imitation rather than inherited.

Components of Personality

Thus far we have been talking about personality as a whole, with little mention of its specific components. Although we have not defined all aspects of the concept personality, there are several traits we relate to it, which are present in all individuals and which are integrated into the total system at different points during childhood. Among these are aggression, achievement, autonomy and attachment–dependency. Two of these, attachment–dependency and autonomy are areas of particular development during infancy. Attachment, described at some length in the text, begins at birth; autonomy, or the sense of self, develops in an important manner during the second year of life. These and other behavioral manifestations of personality will be discussed at greater length in Units 15, 22, and 26.

Study aids

Review Questions

1.   a. Jean likes rock music.
       b. Jean is hot tempered.
          Are both of these descriptive characteristics of personality traits? Why or why not?

2.   Describe how an infant's predisposition toward a certain kind of behavior might shape his environment so as to reinforce that behavior.

3.   Order the following by age and theory:
       a. Phallic stage
       b. Autonomy versus shame, doubt
       c. Oral stage
       d. Anal stage
       e. Imitation
       f. Basic trust versus mistrust

4.   In which of the three theories as they are presented in this unit might toilet training play an important role in the personality development of an infant? In what way does this occur?

5.   What indication do we have that attachment is important to personality development?

Questions Pondered by Psychologists

1.   Early mother–infant interaction seems to affect the infant's future ability to solve problems and to be comfortable in problem-solving situations. Why is this so?

2.   How much of the behavior that comprises personality is genetic or biochemical in origin? There is evidence that schizophrenics inherit a predisposition to the dis-

ease; users of amphetamines report a kind of personality change in response to the drug.

Of Policy Matters and Public Interest

Should mothers of young children work? Opponents hold that the mother should be with a young child, citing studies concerning attachment and bonding. Proponents point out that there is no evidence that the mother must be with an infant full time or what the significance of other caretakers—a good babysitter, a grandmother—might be. They also contend that the quality of interaction is important and that for many mothers the quality would decrease significantly if they had no other outlet for energies and interest. Opponents rely that working mothers are often too tired to bring any quality to their interaction with children after work.

10

Individual Differences

Student objectives

1. Cite some of the purposes and appropriate and inappropriate uses of infant tests.

2. Discuss the significance of sex differences in this period.

3. Describe some of the effects of various child-rearing practices on language acquisition, conceptual development, and personality development as they relate to individual differences.

4. Discuss some of the influences of social class and ethnicity on the child.

Assignments for this unit

1. Review pages 228–233 and pages 244–246. Read pages 196–211 in *A Child's World*.

2. View Program 10, "Individual Differences."

3.   Read Overview 10 and Social Class and Individual Differences in this book.

4.   Review the study aids.

Overview 10

An elderly gentleman who dislikes children developed what he felt was an appropriate social response when he was called upon to admire infants. Each time he was presented with a squalling bundle by its proud mother he would say, "Well, that *is* a baby!"

It worked beautifully for each mother is certain that her baby *is* a baby—something special and unique. It may be necessary to put an identification bracelet on an infant in the hospital, but she *knows* which one is hers.

Many individual differences, no matter how slight they may be to anyone but a baby's parents, manifest themselves at birth and continue to be developed thereafter. Both genetic and environmental factors interact to foster them, and they increase in scope and quantity as the child grows.

Researchers have examined young infants by a variety of tests. They want to describe the ways in which infants differ and to ascertain the range of variation. One result of statistical analysis of these tests has been the identification of areas in which individual differences among babies can generally be found: gross and fine motor development, social responsiveness, goal directedness, and vocalization and language.

An analysis of differences in these activities and the patterns of development that follow them can be approached on several levels. The first raises the question of how large the differences are. The second asks what these differences imply for the child's future, and the third asks how the differences originated.

Simple measurement of differences is obtained by objective observations, free of theoretical presuppositions or conclusions. How do children differ in each dimension studied? This is the type of groundwork needed to furnish an empirical base for further work.

Considerable variation is found in the ages at which infants acquire new skills or attain well-marked stages. One child, for example, may begin to walk unassisted at ten months, whereas another will wait for two years—twice the lifetime of the early walker. It is also difficult to predict whether differences that appear early in life among children will continue as the children grow older. Most children will probably have the same walking ability at age four, but early verbal differences may continue to exist. A child's IQ assessed when the child is six months often bears little resemblance to that same child's IQ at the age of two. The appearance of one skill at an early age might suggest genius to fond parents; the development of this particular skill may plateau slightly later while other abilities catch up.

As the types of differences are established, psychologists can study how these differences will affect the child's future. Some differences—verbal capability or mechanical aptitude—will eventually be important in the choice of a career. Others, such as skin color, sex, or personality, may be instrumental in determining social behavior. Taken together, people's individual characteristics affect how they perceive themselves and how others perceive them.

Psychologists are also interested in the cause of differences. To what extent are differences in mental style or behavior a result of upbringing or of hereditary factors? To what extent is a particular trait stable or transient? Are stable behavior patterns the result of early conditioning, steady and lengthy training, or inheritance? Why is one child highly verbal and another unable to use language to its fullest advantage?

Importance of Individual Differences

Not all early differences are equally important in predicting later personality development or intellectual attainment. One characteristic that does seem crucial is curiosity. Some infants seem to be curious about their environment from birth. They tend to search out new experiences and to explore; other children are passive; still others are hyperactive and randomly restless. It is quite possible that purposefully active infants will increase their competence and confidence in dealing with their environment more quickly than will children who do not explore. It is also possible that early differences of this type can amplify differences in other dimensions. Little is known of the biological basis of confidence or curiosity, but such traits (if, in fact, they exist as stable independent aspects of personality) are likely to be highly influential in determining growth and development.

Erik Erikson's theory of development suggests that an early establishment of trust is vital for all further growth. Infants must feel secure. If Erikson is correct, individual differences in this area may affect all further development. Early failure and deprivation may never be overcome. It *is* true that there seems to be considerable variation and flexibility in responses to stress or success. Many children survive extraordinary hardships with few apparent scars, whereas others are submerged by difficulties that do not seem as great.

Recent observations by Alan Sroufe suggest that mother and infant interaction plays a crucial role in how the young child confronts problem situations. Children who were upset at separation from the mother also performed at a level below average in problem solving; those who seemed trusting and secure in the absence of the mother seemed to bring the same strengths to problem solving. These experiments add a new dimension to the work by Burton White described on page 206 in *A Child's World*, although they do not necessarily supersede it.

The sex of an infant is, of course, an extremely important determinant of his or her future life. Parents and caregivers respond to their children on the basis of sex in many ways, beginning in infancy and continuing through childhood. The information in this unit on sex differences will be continued in Unit 16, "Social Stereotyping" and throughout the units on personality.

The effects of nutritional differences on later development are less clear than those of sex, but severe malnutrition can retard development. This area will be treated at greater length in Unit 12.

Only now are child researchers beginning to explore the roots of individual differences in infancy. At present little is known with any certainty, but the development of new technologies is expected to aid this research. Detailed studies of language and social interaction can be made with the help of tape recorders, videotape, and other technologically based aids, and the next few years should bring out a number of important discoveries.

Study aids

Review Questions

1. Give at least two illustrations of ways in which
   a. A baby's individual characteristics might influence his or her parents' attitudes and behavior.
   b. Parents' characteristics may influence the baby's development.

2. What is the difference between DQ and IQ?

3. What indication do we have that differences in school-age children of different socioeconomic backgrounds are *not* genetic in origin?

4. In experimental situations Baby Jones seemed to want to be with Mother a great deal, showed a slight tendency to choose toys that required fine motor skills, and wore pink booties. Identify Baby Jones' sex and detail the reasons for your choice.

5. If you were a psychologist and a worried father approached you because six-month-old Tony's IQ was not high enough to qualify him for Harvard, what advice might you give him?

6.   What evidence do we have that different child-rearing practices have an impact on the development of individual differences? Draw material from this unit, Unit 9, and Unit 6 for your answer.

Questions Pondered by Psychologists

How does current research about individual differences influence child rearing practices?

Of Policy Matters and Public Interest

In a sense there is no such thing as a "normal" child. We draw averages and look at median ages for the development of different skills, but parents forget that the chances of a child's being right at the average or on the median are slight. Most are above and below. Slightly painful for many parents—who obviously want their child to be perfect in every way—is that average or median implies that 50 percent of all children are below average in any given area and 50 percent are above. A more comforting fact is that few are consistently above or below average. Each child has his or her strengths and weaknesses.

Social Class and Individual Differences

Research material on compensatory educational programs such as Head Start (Unit 18) suggests that the causes of poor performance in school may be established very early in the child's life. Jerome Kagan has identified six areas in which he feels infants from lower socioeconomic backgrounds differ from middle-class infants:

1. *Language.* Poor children neither comprehend nor express language as well as middle-class children. This may be a result of less frequent language interchange with the

mother, or of the type of language the mother uses. Television is not as helpful as direct communication in language learning, and it does not provide a completely satisfactory substitute for "talking" with an adult.

2. *Mental set*. Poor children's environments often do not include the variety and types of interaction that enrich mental structures.

3. *Attachment*. The greater interaction of middle-class mothers with their children enhances the attachment bond.

4. *Inhibition*. Poor children tend to be impulsive; they show less inhibition in times of conflict and spend less time in the consideration of alternatives in a choice situation. This behavior pattern may possibly result from the mother's own impulsiveness or from the child's impoverished mental structures.

5. *Sense of effectiveness*. Lower-class children often seem to feel that they have less control or chance of affecting their environment. This feeling may be a result of the amount of attention and, especially, praise that middle-class mothers give their children compared to the amount lower-class children receive; it may also be a result of the fact that middle-class children have more objects to manipulate in their environment.

6. *Motivation and expectancy of failure*. This appears particularly as the lower-class children enter school, but the factors that bring it about are established much earlier in their life.*

Not everyone would agree with these specific areas but, as both *A Child's World* and the television program illustrate, the environmental differences affecting lower- and middle-class infants do vary and the results are, unfortunately, fairly predictable.

*Jerome Kagan, "On Class Differences and Early Development," in *Education of the Infant and Young Child,* ed. Victor H. Denenberg (New York: Academic Press, 1970).

11

Preschool Physical Development

Student objectives

1. Describe the significance of maturation and readiness as contributors to the development of mental, physical, and emotional capacities.

2. Describe the characteristics of physical and motor development of preschool children.

Assignments for this unit

1. Read Chapter 7 (pages 259–271) in *A Child's World*.

2. View Program 11, "Preschool Physical Development."

3. Read Overview 11 and review the study aids in this book. Although the material is important, this is a short unit. Take a breather.

Overview 11

We generally characterize the years from birth to age three as infancy and from three to six as preschool. The division between the two periods in real life is not nearly as clear-cut. Growth and development occur on a continuing basis, and the baby of yesterday does not suddenly become the preschool child of today. At age three, most children can walk, talk, and use symbols in thought. In none of these activities has full development been reached. Each will be refined and changed throughout life. Even in this short, preschool period, children of six will have moved far from their three-year-old selves.

The basis for much of the change that occurs in the preschool years is physical. Children grow in height and weight and increase in motor skills. The manipulation of objects improves. From awkward circle drawers at three, children at six become accomplished pencil holders, ready to print their name. Gross body movements change in the same way; children graduate from tricycles to bicycles.

Many things affect preschool children's physical growth. Nutrition (explored further in Unit 12), emotional relationships (see the text and the program), heredity and general environment. It is easier to see the sum total of their effects than to measure the contribution of each. A useful general indicator of motor development is body posture. Posture usually refers to the way the body is held while standing, sleeping, or sitting. "Good" posture implies a well-balanced muscular system and is a precondition to well-executed movement in school, at home or wherever a child is physically active.

In this period a new element enters the ability to perform physical tasks. During infancy most new physical behaviors are a result of *maturation*. Maturation refers to the coordinated changes in the body's physical system (Unit 6). Muscles become larger and stronger, cartilage is replaced by bone, and the brain develops new patterns of structural organization through interconnections. As these developments occur, children are able to bring all these systems to bear on new tasks, such as walking. Maturation proceeds through stages, and at each stage children are able to perform new types of action. These physical activities have been called *autogenous*, or self-initiating, and appear quite independently of training, although they may have to be practiced before they are perfect. *Maturation* cannot normally be accelerated, but it can be impeded by many of the factors that impede physical growth.

But the ability to *do* is not the only capacity that matures. The ability to learn to do, or *readiness*, appears in the preschool years. In a sense it is also a function of maturation, but the capacity to perform is no longer *autogenous*. The child may be ready to master new skills, concepts, values, and social relationships; but the learning will not take place unless there is an opportunity to learn. These kinds of opportunities, of course, are abundant in the lives of most preschool children, but they can be enhanced.

Maturation and through maturation, readiness, are prerequisites to the performance of many kinds of activities. Children cannot learn to ride a bicycle before they can control certain muscle sequences and balance themselves on a moving vehicle. Reading cannot be taught to children who are not able to manipulate symbols or coordinate their eye movements back and forth across a page.

As the text points out, physical growth can affect self-image and personality development. Maturation and readiness also have an effect on emotional development as well as on mental and physical growth. If inappropriate tasks are required of children, their inability to perform—often accompanied by adult disapproval—can be the source of later problems. The level of maturation of children and the expectations of the adults around them interact with self-image. All the aspects of a child's development are related, and if one is affected, either positively or negatively, the others are likely to be affected as well.

Study aids

Review Questions

1. Name three environmental factors that can affect growth.

2.   If David were 39 inches tall and weighed 44 pounds at age four, would he be most likely a(n)
   a. Ectomorph
   b. Mesomorph
   c. Endomorph
   How did you arrive at this conclusion?

3.   Is maturation determined by readiness or readiness determined by maturation? Explain your answer.

4.   In contrast to children from underprivileged homes, children from more advantaged home situations
   a. Experience sexual maturity earlier
   b. Have teeth erupt sooner
   c. Grow taller
   d. Attain their full height earlier
   e. All of the above

Questions Pondered by Psychologists

Why, when, and how does a child move from gaining the ability to *do* through maturation to gaining the ability to *learn to do*? What kind of relationship exists between personality type and body type? Does the build reflect a personality, shape it, both, or neither?

12

Nutrition

Student objectives

1. Identify possible effects of nutrition on physical, mental, and personality development in children up to three years.

2. Identify the basic nutrients needed for growth.

3. Identify recommendations for improving diets of growing children.

Assignments for this unit

1. There is no specific text assignment for this unit, but related material may be found on pages 51–54, 130–132, and 204–205.

2. View Program 12, "Nutrition."

3. Read Overview 12 in this book and review the study aids.

Why is it necessary to include a program on nutrition in a child development course? It is possible to argue that as long as children get enough to eat, they will grow up normally and remain healthy. Unfortunately, however, "enough food" is not the same as an "adequate diet," and a large number of children do not get enough food.

In many countries throughout the world, food is scarce for most of the people most of the time. When crops fail, the scarcity becomes famine; even if crops are abundant, the children may not be well nourished. To judge the quality of a diet, we must distinguish between a diet that provides adequate calories for energy and a diet that is also nutritionally balanced. The former diet is inadequate for growth and health, though its deficiencies may not always be as obvious as the simple lack of food. It is ironic but true that malnutrition can exist even where food is plentiful because eating habits may exclude essential ingredients from the diet. In the United States several million children and adults, for instance, suffer from malnutrition and dietary imbalances serious enough to cause harm. A sizable portion of those are financially able to purchase the food they need for a balanced diet.

What *is* an adequate diet for children? This question is not easy to answer. The science of nutrition is relatively new and there are few firm conclusions. It is difficult to do experiments with foods under normal conditions for we consider it unethical to give children a deliberately deficient diet. Scientists can look at the biochemistry of food ingredients in order to judge their effects on the human organism, but they cannot always be sure that their conclusions are correct because of the complexity of metabolic systems.

An adequate diet is difficult both to determine and to implement because food must not only be available but be agreeable. Babies and young children often have strong food preferences regardless of nutritional values. Adults also have likes and dislikes that are based on familiarity and cultural norms. Many Americans don't like snails, a French delicacy.

High-yielding varieties of rice have been rejected by many Asian farmers because of the taste; people are willing to pay less for them than for the "old" varieties. Some African tribes eat the green algae skimmed off stagnant ponds—a nutritionally wise practice for it is rich in protein—but one that North Americans find generally unacceptable.

A further complication is that humans need certain dietary ingredients at the same time if they are to be effective. Vitamin D, for example, should accompany the ingestion of calcium or the calcium will not be digested. This means that a calcium-rich diet can still result in a deficiency in calcium unless there is the simultaneous presence of vitamin D.

Diet complexity is compounded by the advertising claims of food companies, competing claims by food faddists, and conflicting medical opinions, which affect personal choice in eating habits. Many respectable scientists work for food companies and lend their authority to dubious food claims. Everybody seems to be able to say something about nutrition, and many of the statements are contradictory.

Finally, everyone is different. Some people need more of one thing than another. Women, particularly if pregnant or nursing, need more iron than other people. Fast-growing children and those adults who do a lot of physical work need more protein and calories. There is even some evidence that an individual's body adapts to the food of its culture. Metabolic processes may relate differently to the same foods in people from backgrounds with different eating habits.

Good nutrition is important at all stages of life. A wealth of statistics links diseases and health problems to poor diet, especially among children and older people. Yet, as we have seen, the identification and provision of adequate diets for all people may be a complex and difficult task.

Malnutrition in the Early Years

Infants are particularly vulnerable to the effects of poor diet. They grow rapidly and require a steady intake of calories and essential nutrients in order to fulfill their genetic "blueprints." Physical and mental stunting can be the result of an inadequate food supply in the early years. The brain, for example, nearly triples in size during the first year after birth. Brain weight increases from an average of 340 grams to 970 grams in the first year. By the age of three, a child's brain has developed to about 80 percent of its adult weight. If the child is inadequately nourished, this growth is inhibited.

Mental development reflects the significance of brain growth during this period. As we have seen, during the first three years, children learn to walk, talk, and carry on a number of activities—a tremendous step toward adulthood. Malnutrition in these years has severe and often permanent consequences for intellectual growth. Unfortunately, it is difficult to detect malnourishment in the early months since the effects only become obvious when they appear later as mental deficiency. In studies conducted in developing countries, researchers have seen the effects of specific dietary deficiencies and the age of the child at the time of nutritional deprivation (the younger, the more severe) on this type of mental loss. The insidious effects of poor diet are illustrated here—they can be as serious as outright starvation, but much less obvious.

Authorities agree that breast milk is the ideal diet for infants. It contains sufficient calories, vitamins, and minerals for the growing child. In addition, statistics indicate that breast-fed infants are less susceptible to disease than bottle-fed children. This is because many of the mother's antibodies against disease are in her milk. They build the infant's resistance to infection by bacteria or viruses, including polio and malaria. The lower disease rate is also related to some extent to the higher chance of infection associated with improperly sterilized bottles and formula.

Nutrition in the Womb

In a sense, concern with the infant's nutrition begins too late; nutrition begins before birth (Unit 3), but there is an intermediary between the fetus and the outside world—the mother.

After implantation in the wall of the uterus and the formation of the placenta, the embryo requires nutrients to sustain its rapid growth. A poorly nourished mother may not develop an adequate placenta during this important period of development. Each day a normal placenta passes about 300 quarts of blood, which carries nutrients to the fetus. A deficient placenta will pass fewer nutrients and restrict fetal growth.

The critical period for the growth of brain cells (neurons) during prenatal development, for instance, seems to lie in the first 15 to 20 weeks of pregnancy. By the end of the second trimester, the fetus possesses its full complement of brain cells. Development after this time is expressed as cell growth—the brain cells grow in size and complexity, not in numbers. Experiments with animals have shown that those born of mothers with deficient placentae had fewer brain cells than normal newborns.

Protein is especially needed by the prenatal infant. A study of poor black mothers showed that birth weight and skull volume of their newborns was related to the amount of protein in the mother's blood stream. The less protein, the smaller the total weight and skull volume.

A conflicting study was done of children of Dutch mothers who suffered from malnutrition during World War II. Boys born during this period showed no observable differences from older and younger draftees when they were tested for the Dutch army at the age of 19, although their birth weight had been 7 percent below normal. This may have several different implications. Mothers' malnutrition may have to reach a critical point before it affects children—the Dutch mothers may have had reserves built up through years of an adequate diet that are not present in chronically malnourished mothers. It may also indicate that this degree of deficiency in weight at birth was not great enough to overcome the metabolism's compensating mechanisms. Furthermore, we must remember that infants born of chronically malnourished mothers also tend to be malnourished after birth as well as before, thus clouding statistics about critical periods for fetal and infant nourishment.

Clearer evidence that birth weight and intellectual ability are related can be obtained from the study of identical twins. These twins may be born of slightly different sizes even though they are genetically identical because of differences in the intrauterine environment. Dr. S. Gordon Babson has followed identical twins, who were born of markedly different sizes, over a number of years (one twin at least 25 percent smaller than the other with a birth weight of less than 2000 g or 4 pounds 7 ounces). He found that the smaller twin was slower to develop, physically smaller, and exhibited a lower IQ than the larger twin. This finding is also true of most newborns who are underweight for their conceptual age (as opposed to premature babies whose conceptual age is below normal for newborns but whose weight may be normal for their conceptual age).

Basic and Necessary Nutrients

Nutritionists divide the food ingredients necessary for life into five classes: carbohydrates, fats (or *lipids*), proteins, vitamins, and minerals. The first three provide energy for the body, although proteins have to be converted into carbohydrate form before we use them for energy. Vitamins and minerals are also essential, although they are only required in minute quantities.

Proteins represent the essential chemical constituents of biological structures. They combine with lipids to create membranes that preserve the integrity of cells. In addition, the enzymes that direct all biochemical reactions are proteins. Proteins are formed by combinations of 21 amino acids, often in chains of thousands or millions of these submolecules, and are available in all organic life in infinite variety. The digestion of protein is accomplished by breaking these complex molecular structures into their constituent amino acids. The body then uses these amino acids by recombining them to make other proteins that are more suitable to the organism.

Proteins cannot be made from carbohydrates or lipids, but only from other proteins, hence, the importance of ensuring adequate protein in the diet. The body continually renews its proteins and, in order to do so, must have a supply of amino acids. Some of these can be obtained by converting one type into another, but there are eight amino acids that cannot be manufactured by the human body. These must be supplied by food and all at the same time in order to be absorbed. If even one of the eight is omitted from a meal, the effectiveness of the others is severely reduced. This is partly why protein deficiency is a major problem in the world today. Balance in a diet is extremely important.

Carbohydrates provide the major source of energy for the muscles and the internal body systems. They are stored in the liver and muscles as *glycogen,* or animal starch. *Fats* serve in building cell membranes, but they can also be converted to carbohydrates, when necessary, through an elaborate chain of biochemical reactions.

Vitamins cannot be converted to energy nor can they serve as building material, yet they play a vital role in the biochemical reactions of the body. Vitamin A deficiency leads to blindness. Pellagra results from vitamin B deficiency and was common in many parts of the U.S. until recently. Scurvy, a vitamin C deficiency, was one of the major hazards of early ocean travel. Vitamin C is found in fresh fruit, especially citrus fruits and tomatoes, as well as in fresh vegetables. Vitamin D is present in eggs, liver, butter and cod liver oil, but, for most people, is available from sunlight. Rickets, resulting from vitamin D deficiency, was prevalent in the heavily polluted mill towns of nineteenth century England, in Middle Eastern harems, and in other places where children and adults neither received enough exposure to the sun nor to the foods containing this substance.

The essential minerals are chiefly sodium, potassium, calcium, magnesium, iron, copper, iodine, chlorine, phosphorus, sulphur, and fluorine. Sodium, potassium, calcium, and chlorine are common ingredients of most body fluids inside and outside the cells. Calcium, in addition, builds bones when vitamin D is present. Iron is the essential ingredient in hemoglobin in the red blood corpuscles. Iodine is necessary in the manufacture of thyroxine, which is secreted by the thyroid gland into the bloodstream. Thyroxine controls body growth, and its lack produces dwarfism and mental deficiency. In most areas sufficient iodine is naturally available in the drinking water, and table salt is often supplemented with it.

Eating Habits

Most children can be fed adequate and nutritive diets with little trouble today. Nutritionists, home economists, and dietitians provide continual advice about selecting balanced diets for the family; most Americans get the foods they need in the course of a normal day. Eating habits, however, can make the difference between good and bad health. These include the type of food eaten as well as when and how it is consumed. Tough food needs to be chewed well. Infants and young children should have a minimum of foods that are hard to digest, and older adults should generally avoid an excess of carbohydrates. Meals should be frequent—three or four a day—and small rather than few and large to avoid overeating at any one sitting.

The cooking of food is an important factor in nutrition. Many vitamins are lost by excessive boiling, but a certain amount of cooking is often necessary in order to make foods palatable. Cooking can eliminate harmful bacteria and other microorganisms in foods as well as destroy some types of parasites, such as tapeworms. Almost all human societies cook a large portion of their food.

These are all aspects of nutrition, a field that represents many disciplines: biochemistry, biology, medicine, psychology, sociology, agriculture, and, in today's world, politics. It is becoming more important to distinguish fact from fad. What we discover about nutrition now and in the future can affect the lives of millions.

Review Questions

1. If a child spends a long, dark winter indoors and eats absolutely no meat, dairy, or fish products, what vitamin deficiency is likely to occur?

2. Infant X does not yet have all his brain cells. Infant Y's brain has just reached about 80 percent of its adult size. What does this tell you about their approximate ages?

3. What conflicting information do we have concerning the relationship of nutrition and mental development in infancy? Can you suggest possible ways of reconciling this?

4. What are the nutritional advantages of breast-feeding?

5. What is a necessary ingredient in the formation of protein?

6. What should the diets of growing children contain?

Questions Pondered by Psychologists, Anthropologists, and Nutritionists

Why are all foods not equally acceptable to all people? Do religious prohibitions have a health or ecological basis, or did people give up what they originally had liked best? Why do some people have a "sweet tooth" and not others?

Of Policy Matters and Public Interest

We have established standards of good nutrition based on the foods we eat and those that work well for us. We have also extended these standards to other areas and, by using them, have found that most people do not have enough food or nutritionally adequate diets.

Recently some researchers have pointed out that millions of individuals survive quite well on diets that contain fewer than the minimum number of calories necessary for health by World Health Organization standards. They also point to statistics in various areas that simultaneously show that (1) per capita caloric consumption is decreasing and (2) life expectancy is increasing. These researchers then ask if the statistics or minimum standards for a "nutritionally adequate diet" are really accurate.

13

Preschool Mental Development

Student objectives

1. Describe the characteristics of the child's thinking.

2. Explain the developmental (Piagetian) and behavioristic theoretical approaches to the preschool child's mental development.

3. Describe possible effects of child-rearing techniques on mental development.

Assignments for this unit

1. Read pages 273–285 in *A Child's World*.

2. View Program 13, "Preschool Mental Development."

3. Read Overview 13 in this book and review the study aids.

4. Read A Perceptual Approach to Mental Development in this chapter (optional).

Infants learn the constancies or *invariants* of the world immediately around them during the first year of life. When visible objects go out of sight, they also, as far as infants are concerned, cease to exist. With experience, children come to perceive objects as permanent whether they can see the objects or not.

Children will also sort out the unchanging aspects of common *events* during the next few years. The consequences of this achievement are significant. A child who anticipates the outcome of an action ("If I go into a dark room, a monster will eat me up.") demonstrates the ability to think about the action without support from the environment (being eaten up or, in this case, even seeing the monster). The widening powers of children's minds ultimately lead them to knowledge concerning events in the past and the ability to imagine the future.

This transition illustrates some of the growing mental capabilities of preschool children. Learning to use symbols is one of the most important. By the time she is three, Carla can think about her mother when she is away. Carla can also exclaim "Safeway!" whenever she sees the red S, whether it is on a store or a can.

As a part of the symbolic function—in the sense that words are symbols—Carla has also gained communicability. Throughout the preschool years, she will be able to communicate her wishes, thoughts, desires, and fears with increasing fluency. Correspondingly, it will become easier for her to learn through the medium of words as time goes on.

Preschool children learn some of the regularities and correlations of the real world. When Jimmy kicks a ball, the ball moves. When the sun goes down, it is dark unless we turn on the light. Many of the perceptions relating to causality and the succession of events are still faulty, but the kernel of the idea exists. Other concepts—of space and time and identity, for instance—appear in the same form. They exist but may still be flawed. Their greatest development comes at the end of the preschool period and will be discussed in Unit 19.

PRESCHOOL MENTAL DEVELOPMENT

Piagetian Model

Piaget calls these preschool years—those between the ages of two and seven—the *preoperational stage* and characterizes children in this period as exhibiting certain qualitatively distinct modes of thinking.

Egocentrism

A young child still cannot see the world from the point of view of another person. Drawing pictures illustrates an aspect of this phenomenon. Children do not begin to use perspective in pictures until they are around eight to ten years of age. Before that time everything is flat, and there is no consistent use of depth. The use of perspective represents awareness of how things look from a single viewpoint. During the preoperational stage, children have little awareness that they have unique points of view, either in visual or in social perspective. Visual egocentrism generally disappears with the advent of the stage of concrete operations, but many would argue that most adults retain at least some social egocentrism throughout their lives.

Centration

Preoperational children tend to focus or *center* only on one aspect of a problem or situation. If there are two or more aspects that need to be taken into account, such as height and width, children make "mistakes" when presented with problems. One such problem that has been the subject of many experiments involves numbers.

Children appear to handle numbers in the same manner as people may have done for centuries. This is illustrated by their comparison of groups of objects. It is possible for them to tell which of two groups has more objects if they look for a correspondence on a one-to-one basis.

```
x     x     x     x
o     o     o     o
```

There are an equal number of x's and o's. A more sophisticated person would also count the objects in each group and observe the fact that each group has "four." Note that numeration, or counting, defines the groups as being equivalent without making it necessary to see both of them at the same time.

Before children have properly grasped formal numeration and arithmetical operations, they can tell whether one group has fewer, more, or the same number of objects if they see them together. But if the experimenter, even in plain view, spreads one group apart so that they look like this,

```
   x   x   x   x

    o   o   o   o
```

a four- or five-year-old will say that the top row has "more" in it. Alternatively, if more objects are put in one row though the lengths remain equal,

```
   x     x     x     x     x     x

   o   o   o                       o
```

the child will tell the experimenter that the number of objects in each row is "the same."

The child's difficulty here is characteristic of his attempts to analyze complex situations. He fails to interpret—or perceive—the display correctly because he pays attention to only one feature—the length of the rows—while disregarding other aspects. When most children are six or seven years old they learn to *decenter* or pay attention to more than one aspect of some situation. In other situations, however, decentering comes considerably later.

Focus on States

A related point is that young children are not capable of dealing with certain kinds of manipulations or changes in materials, even though they judge static or unmoving displays correctly. At around five or six years of age, Henry might be able to put as many pieces of gum in one row as in another. If the experimenter moves the pieces of gum in one row close together, Henry will still claim that the other row has more. Piaget suggests that this type of answer arises from an incapacity to process "kinetic" or moving displays. Henry is not yet ready to handle that much information at one time.

Transductive Reasoning

At this stage the child's reasoning illustrates another aspect of conceptual immaturity. Inconsistency of thought is common since the same object or event can be classified differently on different occasions. Both *inductive* and *deductive* reasoning are often absent, leaving what Piaget calls *transductive* reasoning.

Deductive reasoning, in Piaget's terminology, means the ability to form correct subordinate categories. If you were asked to name subordinate categories of animals, you might name dogs, cats, and horses. Inductive reasoning is the reverse; given the categories dogs, cats, or horses, you should be able to name the superordinate category, animal. *Transductive* reasoning leaps across these orderly hierarchies. A three-year-old can ignore the principle of *class inclusion* (a tiger and a lion are both animals) to claim that one is an animal and the other one is a tiger and not an animal. An older preoperational child, presented with four figures of lions and three of tigers, might just as seriously answer that there are more lions than there are animals, even though she knows lions and tigers are both animals. The complexities of manipulating different levels of relationships are clearly greater than most adults remember.

Imitation

According to Piaget, sensorimotor *imitation* comes before thought in this period. The child must create some kind of mental representation in order to think about objects and events that are not present. With imitation the child develops an internal model, which eventually replaces imitative physical actions such as "bye-bye." Through this imitation three broad types of mature representation eventually emerge from the preschool period: concepts, symbols and images, and linguisitic signs. These can represent concrete objects or events, as in a mental image or symbol, or they can be abstract, as in a *concept*. Linguistic symbols come to provide social signs that do not resemble the thing to which a reference is made.

Behavioristic Model

Unlike Piaget, behaviorists believe that events in the individual's environment influence learning. Robert Gagné, for example, feels there are eight *conditions of learning* or *learning types*. These are (1) signal learning (or classical Pavlovian conditioning), (2) stimulus-response learning, (3) chaining of individual stimulus–response actions, (4) verbal association or chaining, (5) discrimination learning or learning to make a number of different responses to different stimuli that may resemble one another in physical appearance, (6) concrete concept learning—grouping concrete objects so that a response can be made to them as a class, (7) rule learning, and (8) problem solving.

Behaviorists do not believe that children must pass through predictable, progressive stages of learning, although earlier conditions of learning must generally be present before later ones can operate. Behaviorists feel that as children learn, they are engaged in the process of building these structures of learning in many different areas simultaneously.

Whether theorists subscribe to the belief that learning in preschool children follows an inwardly determined path or that it occurs primarily as a result of environmental input, most agree that the environment is important in mental development. The exact role it plays and the precise manner in which it operates, however, is much less easy to determine.

Curious, alert, assertive children generally learn more than passive, apathetic, withdrawn children. What causes each type of condition? Clearly, a certain amount of stimulation from the environment is necessary, but too much or too disturbing stimulation can possibly hinder learning. Children raised in environments in which there are many books and in which reading is a valued activity tend to learn to read earlier than children who have had no such exposure. Since many learning tasks (for example, writing) require basic skills (holding a pencil), preschool children who have had an opportunity to practice these skills may have an advantage compared to those who haven't. That does not, however, preclude the possibility that a child who has not had such a chance can catch up.

Some studies have been done that focus on the interaction between mothers and children as such interaction relates to the children's mental growth. Burton White, a psychologist at Harvard University, found that mothers' attitudes and temperaments made a significant difference in the development of their children. Mothers who interacted with their children in a positive, helpful manner at points when the children particularly needed them, who were energetic, cheerful, and relatively tolerant of minor risks and small messes, had children who generally were rated as excellent. Mothers who regarded children primarily as a burden, who were overprotective, who were unable to organize their homes, and who found it difficult to cope with life, had children who generally did not score as well. A fuller account of White's work appears on pages 206 to 208 of the text. This section will be assigned in the unit on intelligence testing.

Training can aid children in specific Piagetian tasks, especially if they are moving into a transitional period toward the point where they would learn the skills naturally. If children are not close to this transition period, however, they may "learn" how to do one of the tasks described in this unit, but that ability does not generalize to similar tasks using different kinds of materials. Furthermore, the effects of training are generally short-lived.

Study aids

Review Questions

1. In which approach to mental development—behaviorist or Piagetian—does environment play a greater role in the preschool years?

2. All of the following are signs. Why?
   a. Tree
   b. Cat
   c. π (pi)
   Which of them could also be symbols in a child's imagination?

3. The above, rather difficult question contains a class-inclusion problem. (All are signs because they are all words; a child could picture—or have symbols for—tree and cat, but pi is still an arbitrary representation of the mathematical concept.) What was it? Why would you expect a child in Piaget's preoperational stage not to be able to solve it?

4. Is there any evidence that mothers' attitudes affect children's mental development? How?

5. "He's wearing long hair and has a beard so he must be intelligent." What kind of reasoning is illustrated by that statement?

6. What is *centration*?

7. Identify two of the major differences between thinking in infancy and thinking in the preschool child.

8. The earliest or lowest level of classification ability is called:
   a. Quasi classification
   b. Class inclusion
   c. Figural collections
   d. Exhaustive classification
   e. Exhausting classification

Questions Pondered by Psychologists

Is the preschool child's growing ability to render lifelike drawings a result of perceptual, cognitive, or motor development? Is it possible to think without using symbols or signs?

A Perceptual Approach to Mental Development

There are a number of theoretical approaches to mental development besides those of Piaget and the behaviorists. The approach to perceptual development proposed by James and Eleanor Gibson, for instance, represents a view that does not involve either the notion of mental *schemata* or S–R links. The Gibsons propose that learning—perceptual or cognitive—proceeds on the basis of *differentiation*. The child does not learn to associate stimuli and responses (as the behaviorists believe), nor does the child have to form sensorimotor schemata through activities (as Piaget believes). Instead, progress in mental development simply reflects better discrimination. Infants' motor behavior is useful in detecting differences between objects and events, but they do not have to use it to form mental structures representing these differences. Where Piaget would claim that the child develops a concept or schema of a *permanent object* (see Chapter 7), the Gibsons would say that the child simply learns to *distinguish* different kinds of events. Presumably, when an infant can perceive that an object has been hidden, she will then search for it. It may be unnecessary to say that a *concept* (that all objects are permanent) is formed.

14

Developing Language Skills

Student objectives

1. Discuss the relationship between language and thought.

2. Identify factors influencing the development of language (child-rearing practices and social class and culture).

3. Contrast egocentric and socialized speech.

Assignments for this unit

1. Read pages 296–307 in *A Child's World*.

2. View Program 14, "Developing Language Skills."

3. Read Overview 14 in this book and review the study aids.

4. Read Language and Culture in this chapter (optional).

Zia never seems to stop talking. She sings to herself in the bathtub, asks for candy, and demands Baby, Lambie, and Raccoon before going to bed. She answers questions about her day when she feels like it (or, perhaps, when she knows the words), and she talks to her toys as she plays with them.

Zia, like other children at age three, has a language capability greater than any known animal besides humans, but her speech is still qualitatively different than that of most adults. Her vocabulary is relatively small, she makes mistakes in syntax, and much of her conversation is *egocentric*. By the time she leaves the preschool period, however, she will have tripled or quadrupled her vocabulary; she will use complex forms in the correct syntax; and she will have moved to the use of *socialized* speech.

Much of this development in language parallels the development in cognitive ability discussed in the last unit. The close relationship of language and thought raises a number of questions: Does language determine thought or is language determined by mental development? Is a certain level of thinking necessary to the ability to speak? Does the language that is learned enable a person to think, using words to describe concepts and relations?

Piaget and a Russian psychologist, L.I. Vygotsky approach these questions from opposite directions. Piaget is, perhaps, the best known supporter of the school that claims that language development reflects the level of cognitive development. He holds that children's ability to learn words and grammatical relations depends on their perceptual and cognitive differentiation of objects and their relations to the world. In other words, children must know *about* something before they can use words to *describe* it. Language plays little or no part in the unfolding cognitive capabilities of the child. Language is, instead, a reflection of the various stages of mental development discussed in Chapters 7 and 13.

L. I. Vygotsky, on the other hand, holds that language strongly influences cognitive development. For both adults and children, speaking goes beyond the expression of thought to the creation of it.

Both Vygotsky and Piaget agree on the general characteristics of preschool children's speech, although their interpretation of them differs. Piaget's original observations were published in 1926. In this book Piaget described the speech characteristics of young boys and girls based on close observation of children and their interactions. He found that they did not always use speech to communicate with one another. Piaget called this *egocentric speech* and divided it into three categories or types.

1. *Repetition.* Often children will repeat words or phrases, almost echoing the words of the speaker. This *echolalic* activity seems to give children pleasure but carries no intention of conveying information or demand.

2. *The monologue.* In this form of speech children talk to themselves.

3. *The collective monologue.* A group of two or more children carry on monologues. They may glance at one another and, perhaps, be talking of related subjects, but they are not holding a conversation.

As children grow older, they begin to use *socialized speech*. Through questions and answers, threats, commands, and requests, language becomes a tool to interact with other people. The use of *adapted information*—saying something that is primarily of interest to a listener in order to attract the listener's attention—occurs in greater measure around five years of age. Through the use of all of these forms, children change their language from a self-addressing system to a tool to affect the behavior of others, and as these forms come into use, egocentric speech diminishes.

Vygotsky, like Piaget, felt that language was preceded by a prelinguistic phase of thought. He also recognized the existence of *egocentric* speech, or speech not intended for potential listeners. The function of egocentric speech in Vygotsky's theory, however, is different. Rather than being an expression of egocentric thought, Vygotsky's egocentric or *private* speech guides children's behavior and thinking. In monologues, for instance, children are not describing their actions or thoughts but are directing them.

Vygotsky also claims that egocentric speech is not replaced by social speech intended for a listener. Instead egocentric speech increases along with social speech until the child is about six years old. After that time, it declines in observable usage, but not because it no longer exists. It disappears because society frowns on "talking to oneself." Egocentric speech is thus transformed into silent or *inner* speech. It continues to be used through life and its functional value—that of guiding and facilitating thought—remains important because Vygotsky believes that language is necessary to thought.

Both conventional wisdom and scientific evidence exist to support each side of the language–thought controversy. The children's monologue has a parallel among adults. Activities that require problem solving often seem easier with some kind of verbalization, or "talking it out," if only to oneself. This seems to occur even for problems of an essentially nonverbal form, such as in chess, and a considerable body of Russian research indicates that performance in solving certain types of problems deteriorates when private self-guiding speech is disrupted.

On the other hand, it seems logical to presume that we cannot talk about things we cannot think about. One experiment supported this view through an attempt to prove the reverse and make a logical application of Vygotsky's theory.

A psychologist observed that children who had failed to *conserve* quantity (as in the experiments described in Unit 13) explained their failure using absolute terms. "This one is *big*," they would say, rather than, "This one is *bigger*." An older age group, which had moved out of the preoperational stage and had answered the experimenter correctly, explained their answers with the proper relational words.

The psychologist argued that if language played a crucial role in conservation problems, she should be able to teach the preconservers how to conserve by teaching them the proper words. She was unsuccessful. Although she could teach the children to describe a particular demonstration of conservation properly, they could not solve a problem if any of the details were changed. The ability to use language did not, in this case, enable the children to move out of Piaget's preoperational stage in their thinking.

Although we can identify stages of language development and make generalizations concerning the average rate of development, many children do not follow these averages. Much of the deviation seems to smooth out with time. A mother who is anxious because her baby is two and doesn't talk may be worrying needlessly. Some differences in language development, however, are longer lasting and often have far-reaching implications for children's later development. Many differences seem to be influenced by early family relationships and by sociocultural background.

Most children acquire the patterns and habits of speech of the people around them. This is the basis for learning a language, but it is also the basis for many differences in adult speech, whether the differences spring from different languages, dialects, or the use of vocabulary and syntax. Alberta Siegel, a psychologist from Stanford University, has noted that children who are raised in institutions with group care are almost invariably retarded in their language development. Children may need someone with whom they want to communicate. According to Siegel, children who are reared in an emotionally impoverished environment in which there is no one special they care about, their language growth appears to suffer and it may be irremediable.

Very often the person with whom the child wants to interact is a parent. There is some evidence that amount and quality of interaction between young children and their mothers (primarily, one suspects, because they generally spend more time with mothers than with fathers) affects the richness and quality of the language developed by a child.

One widely noted phenomenon is "motherese." People talking to children tend to use simplified grammatical forms and words they feel the child will understand. They often change the volume and pattern of their regular speech, adding special emphasis to some words and slowing down for other phrases. This is a new area of research, and it is not yet known how mothers differ in their speech with children or exactly what the effects are. However, it may aid language development in some manner.

Social class and cultural background affect the development of language in children. The social class differences outlined by Bernstein are discussed in the text. Other researchers have noted that the use of gestures can augment or replace spoken language and that the degree to which they are used varies with class and culture.

Children who are raised speaking two languages (bilingualism) seldom have trouble learning both, although they may learn each one more slowly. Children who are raised using dialects rather than standard speech may have more difficulty than bilingual children in school. For a number of reasons, however, the evidence concerning this is unclear. Although most dialects are as rich in expression and syntax as standard forms of the language, teachers may not accept them as such. If a dialect is considered impoverished speech by a teacher or by people interacting with a child, their attitudes may affect the child's total development. Furthermore, socioeconomic factors are hard to separate from dialect use and are also influential in school performance and language use. On the other hand, some dialect users may blur their use of dialect and standard language forms. It may be that, because of the closeness of the two forms, it is less easy for them to perceive differences in use. Bilingual children, in contrast, rarely seem to encounter much difficulty in keeping their languages separate.

Study aids

Review Questions

1. What is the difference between a *collective monologue* and *socialized speech*?

2. What does overgeneralization of rules indicate about a child's acquisition of grammar?

3. Would a parent from a status-oriented family, according to Bernstein, be more likely to use a *restricted* or an *elaborated code* with a child? What effect might the use of this code have on a child?

4. "One cannot conceive of eternity without knowing the word for it." Would Piaget or Vygotsky be more likely to make that statement? Why?

5. What is *sociolinguistics?*

6. On what observable phenomenon are Piaget's and Vygotsky's descriptions of language development based? How do they differ in their interpretations of this phenomenon?

Questions Pondered by Psychologists, Neurologists, and Anthropologists

Does the mental development of children with speech difficulties differ from that of children without such difficulties? Is there any variation in thought because of language and culture—that is, is there any truth in the Whorfian hypothesis?

Of Policy Issues and Public Interest

Some areas have instituted bilingual education programs because a large number of children in local schools do not speak English at home. Proponents of these programs argue that non-English-speaking children are under an irremediable handicap in schools. They contend that learning basic skills in school is difficult enough without forcing children to learn them in a second language, which has to be learned at the same time. Opponents of bilingual education hold that children will need English to operate in American society. They say that there is convincing evidence that we have difficulty teaching basic skills in English alone and that attempting to convey them in English and another language simultaneously is unwise. They point out such programs often exist in areas where English-speaking children need extra help with standard forms of the language, not half-time education in a second language. Most educators, however, agree that a child's cultural and linguistic background should be respected regardless of the language in which the child is taught.

Language and Culture

Another important aspect of the relation of language and thought is whether linguistic differences are responsible for significant variations in ideology, thought, and behavior. Do Americans and Chinese think differently because their languages are very different? Are their languages different because they think differently?

A strong claim that differences in language affect all aspects of cognitive functioning was made by Benjamin Whorf in 1956. After careful studies of European and non-European skills, social practices, and language, Whorf said that linguistic differences could account for differences in perception, categorization, and thinking.

Under the Whorfian hypothesis, the use of the word *kal* in Hindi could relate to the fact that there has been relatively little writing of history in the traditional dynasties of India. *Kal* means yesterday; it also means tomorrow. It could denote a very different relationship to the passing of time than European words. The concept of time in Sanskrit would support this contention.

There is also a claim for the influence of language on perception. Eskimos possess many different words for snow. English vocabulary is limited to a few broad terms, such as *slush, powder, fresh,* and *packed,* and it is difficult, if not impossible, for English-speaking people to distinguish among very different types of snow the way Eskimos do.

On the other hand, experiments with color perception indicate that the ability to recognize differences in color does not depend on having the words for those colors. One New Guinea tribe has only two words—*light* and *dark*—to describe color. When given tests under experimental conditions, however, tribe members' color discrimination proved to be as good as the average American's.

15

Preschool Personality

Student objectives

1. Discuss the complexity and interrelationship of factors affecting personality and psychosexual development in this stage using the perspectives of psychoanalytic, psychosocial, and social learning theories.

2. Define identification and discuss how it develops according to psychoanalytic and psychosocial points of view.

3. Describe some components of personality in the preschool child including dependency, autonomy, anxiety, and aggression.

4. Discuss how child-rearing practices influence the development of anxiety, aggression, dependency, and autonomy in children.

Assignments for this unit

1. Read pages 333–350 and 363–381 in *A Child's World*.

2. View Program 15, "Preschool Personality."

3. Read Overview 15 in this book and review the study aids. Note that the text coverage of the theoretical perspectives is relatively complete but that the overview adds some new material on objective 3.

4. Read The Roots of Aggression in this chapter (optional).

Overview 15

Children of three can talk with their playmates, argue with their parents, and explore the world around them with inexhaustible curiosity. In all of these activities, their personalities play an important role. Many traits, or behaviors that are relatively stable over time, are becoming more distinct. They shape children's interaction with the world as well as influence the attitudes the world takes toward them.

Many components of the preschoolers' personalities will be present throughout their lives, but other behaviors are related specifically to this stage of development. Preschoolers' self-control over their own actions is much greater than it was when they were infants, but it is not, as yet, the self-control of adults. They no longer grab any object that interests them. Occasionally (if not often) they save a piece of candy "for later," but they still throw sand at other children who make ugly faces at them.

Parents continue to be of particular importance in this period. It is at this stage that children show signs of *imitative social behavior*, or *modeling*. As most parents know, children of this age will repeat words, actions, and gestures—sometimes in awkward and inappropriate situations. *Identification* with the same-sex parent in both *psychoanalytic* and *psychosocial* theory lays the basis for future self-perception and the ability to function in an acceptable sex role.

Playmates and friends are also influential and will become more so during the next few years. A child's desire to interact with other children leads to a gradual diminution of *egocentric* behavior. By the middle years, children will have learned to appreciate other points of view in both conversation and behavior. Successful socialization in the preschool years indicates the child's growing ability to exercise choices and to attain goals in a manner that is acceptable to both adults and other children.

How this occurs and which traits emerge during this period to become lasting influences on the child's personality is a result of a complex interaction of many factors. Biological influences range from body structure and physical abilities to the action of the endocrine glands in influencing activity levels and rate of sexual maturation. Environmental factors include social and cultural influences, levels of acceptance by parents and peers, and organic factors such as illness, injury, malnutrition that cause deviations from normal development.

Theoretical Perspectives

This unit examines two major theoretical perspectives that treat the preschool years as a particular stage in personality development. These are the *psychoanalytic* approach of Sigmund Freud and Erik Erikson's *psychosocial* theory. A third body of theory, that of *social learning* is also applicable during these years. It is discussed in the text in connection with the development of sex-role identity; the social learning approach to general personality development in the preschool years is largely the same as that discussed in Unit 9 on personality development in infancy.

In Freudian psychoanalytic theory the preschool child has passed from the *anal* into the *phallic* stage. During this period the desire by boys for their mothers *(Oedipus complex)* and by girls for their fathers *(Elektra complex)* is gradually overcome as the child *identifies* with the same-sex parent.

According to Erik Erikson's *psychosocial* theory of development, the crisis that occurs during this period is not primarily sexual but has to do with the split between being a child and becoming an adult. Children's *initiative* enable them to undertake, plan, and carry out activities; if they are not allowed to complete tasks or cannot do them, they feel *guilt*.

Autonomy, Anxiety, and Independence

As infants grow into toddlers and preschool children, the nature of the attachment bond (Unit 9) changes. Children develop a desire for autonomy and greater independence from parents. They find delight in mastering new skills. At the same time, however, they need reassurance and encouragement from mothers or caretakers.

The result is an oscillation between self-assertiveness and dependence, almost as if they were both older and younger than their age. Perhaps these rapid changes in their level of emotional maturity reflect their attempts to identify with (and imitate) their parents without having adult control over impulse and behavior. As children gradually make friends outside the family, the dependent bond with their parents is less crucial. The emotional energy invested in the primary relationship (with the mother) is distributed over several relationships. In this way a more stable personality emerges. Children gain confidence in their ability to be independent.

Because autonomy demands self-acceptance, parents can subvert children's growing independence by denying them a secure base for their adventures into the world. An insecure mother, threatened by a child's assertiveness, may withhold the warmth and affection that he needs in order to be assertive. If emotional support is missing, the child can become anxious and fearful.

Lack of warmth and support isn't the only cause of anxiety in the preschool years. A traumatic episode, such as being bitten by a raccoon, can remain so fixed in a child's mind that the incident appears later through adult eccentricities. Sexual incidents such as advances by an adult or teen-ager may lead to strong emotional repressions. Reactions to such anxiety, called *defense mechanisms* by Freud, are discussed in Unit 22.

Aggression

Freud was the first student of human behavior to recognize the devious powers of sexuality. He was so impressed by the pervasive expression of eroticism in his patients that he attributed all human behavior to the sex instinct (*Eros*). Later in life, dismayed by the slaughter of World War I, he came to believe that an aggressive instinct was deeply rooted in humans. Freud called this instinct *Thanatos*, a negative, death-seeking force.

Konrad Lorenz, an ethologist who studied the natural behavior of animals, found support for Freud's view in his observations of lower animals. Lorenz felt aggressive behavior to be universal and, perhaps, innate, therefore inevitable.

Not all scientists agree with Lorenz and Freud, and there is no definitive experimental evidence concerning aggression or of the specific situations that trigger a hostile act.

Assertiveness in preschool children can vary from unacceptable aggressive behavior to maladaptive meekness. A normal child has to learn the limits of assertiveness and to accept constraints through interaction with playmates and adults. This is not always easy; indeed, finding a middle ground between aggression and passivity is a problem encountered by many adults.

Study aids

Review Questions

1. How does *imitation* differ from *identification*?

2. Contrast the *psychoanalytic* and *social learning* approaches to identification.

3. According to Freud, how does the personality development of girls differ from that of boys during this period? Outline a modern reinterpretation of this.

4. Name at least one positive and one negative aspect of the fears that children develop between the ages of two and six.

5. How does dependency change during the preschool period?

6. How might different styles of child-rearing affect aggressive behavior?

Question Pondered by Psychologists

What function does anxiety serve?

Of Policy Matters and Public Interest

Should violence on television be censored? Opponents of censorship cite the American tradition of freedom of expression, point to studies that show little relation to television violence and children's behavior, and emphasize the fact that parents should control television watching. Those who would like to curtail violence on television cite studies that do relate it to children's behavior, point to rising crime rates, and say that it is not always possible to control what children see, particularly if the violence is in a commercial for another television program shown during a "family" program.

The Roots of Aggression

What causes aggression? Why are some people more aggressive than others? Three different—though possibly related—approaches to these questions are suggested.

Two animal behaviorists, Dollard and Miller, have formulated a *frustration–aggression hypothesis*. They propose that aggression increases proportionately with the amount of frustration. This is clearly an adaptive form of behavior—

frustrating obstacles to goal attainment can be overcome through aggressive responses.

Unfortunately, generalized aggressive impulses seem to arise that have little to do with overcoming a specific frustration. The aggressive urge can be displaced to another object—the businessman who lost a contract comes home and yells at his wife. Displacement activity of this type tends to occur more often when the frustration is intense or the path to a goal is clearly blocked.

An environmental influence that is clearly important in the development of aggressive behavior is the parent–child relationship and the family environment. The facts concerning the nature of this influence, however, are not completely clear. Parents who punish their children a great deal tend to have aggressive children. Considerable evidence exists concerning a relationship between violence in the home and juvenile delinquency.

A home in which parents' tempers are short and resentments simmer tends to induce considerable anxiety and dependence in preschool children. This is an age when new competencies are normally developing. Considerable conflict may be generated between the normal urge toward autonomy and the immature anxieties induced by the home situation. Frustration leads to aggression, which is then expressed in ways that invite retaliation from the parents.

Finally, sex differences in the incidence of aggressive behavior are widely reported. Boys display greater aggression than girls in experiments using a variety of different criteria. Some evidence indicates this is a result of genetic differences, but strong social influences on sex roles must also be recognized.

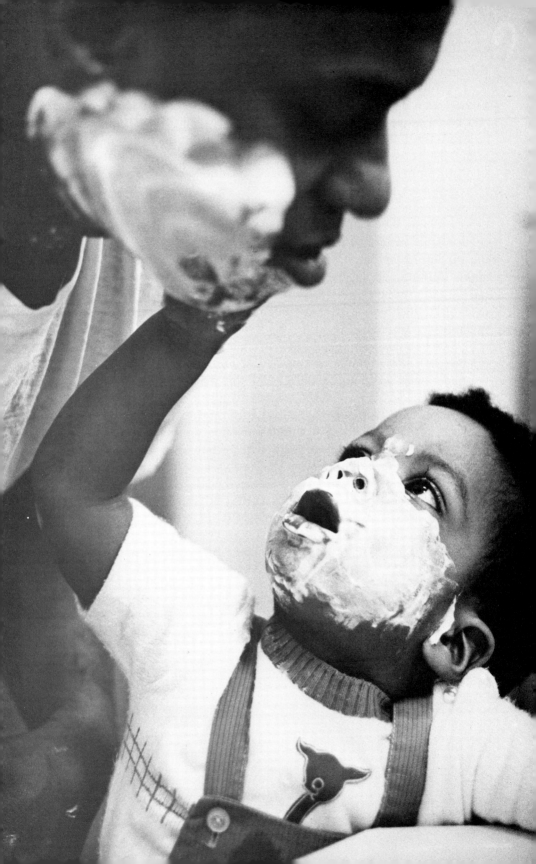

16

Social Stereotyping

Student objectives

1. Describe how stereotyped responses develop.

2. Identify the range of environmental influences on sex-role stereotyping.

3. Discuss three theoretical explanations of sex-role development.

4. Identify some positive and negative aspects of stereotyping.

Assignments for this unit

1. Review pages 228–231 and 343–350 and read pages 350–362 and 512–517 in *A Child's World*.

2. View Program 16, "Social Stereotyping."

3. Read Overview 16 and Sex-Role Stereotyping in this chapter. The overview for this unit covers material that is not in the text.

4. Review the study aids.

Brad waited impatiently for Dr. Morris, listening to the hurried footsteps in the clinic hall. Suddenly a pony-tailed figure skidded around the corner, one hand full of charts, the other pulling absently at a gaping lab coat. "Sorry I'm late," she said breathlessly, adding "I'm Dr. Morris, you must be, uh," she glanced at a chart, "Brad Smith."

If this were a novel, the next line could well read " 'You are Dr. Morris?' Brad asked incredulously." Doctors are supposed to be calm, dignified, in charge of a situation. They are also "supposed" to be male in the stereotype most Americans have. This Dr. Morris violates most expectations—perhaps a good device for a novelist—and we could well imagine Brad feeling apprehensive at the thought of his life and health in her hands. Most of us are more comfortable with familiar stereotypes than with the unknown or culturally inappropriate ones.

Stereotypes, or commonly held beliefs about the characteristics and qualities of people, serve a useful purpose. They provide categories around which implicit knowledge (see Unit 20) can be organized to carry meaning without long explanations. *Truck driver*, *cowboy*, or *librarian* are words that evoke complete images to Americans. Their use reduces the need for further explanation. Stereotypes also carry a certain amount of predictive information. The interests, background, and social position of Dora Hailey, professor of history at the local college are apt to be different than those of Dora Hailey, manager of the local women's softball team. Most people, when introduced to either, would begin a conversation according to their perception of the person holding the position.

Stereotypes of individuals or of roles they are expected to play also provide a framework for behavior. Johnny knows how he's supposed to act when he's at his aunt's for dinner. He also knows the role expectations other children have of him when he's playing baseball. That kind of knowledge is comforting, for Johnny can anticipate expectations and then choose whether to conform to them or to disregard them.

SOCIAL STEREOTYPING

Stereotypes, helpful in many situations, can become destructive when the qualities they portray are negative. When the stereotype of a particular group is not only negative but perceived as the unchanging truth, the consequences can be terrible. Such stereotypes can reduce individual human beings to faceless entities without any of the qualities that would lead "us" to empathize with "them." In this way otherwise decent men and women countenance the extermination of Jews, the lynching of blacks, and the ill treatment of "outsiders."

Children learn stereotyping—with the possible exception of sex-role stereotyping—much as they learn other social conventions. They gain information concerning social roles from their social interactions and the examples set by adults. As they grow old enough for dramatic play (Unit 17) they develop new roles. "If I'm the cowboy, I get the gun. The Indians shoot with bows and arrows." Much of this is a normal part of their socialization into adult roles and behavior; much of it is an aid in ordering the kinds of people who inhabit their world.

There is also evidence that children absorb negative aspects of stereotyping and that they develop prejudices at quite an early age. Studies indicate that American children show an awareness of racial difference as early as age three and can demonstrate negative attitudes toward race soon thereafter.

Children acquire three related types of information concerning social roles and stereotypes.* The first, of course, is simply that there are different social categories in the community and society: men and women; blacks, Mexicans, Germans, and other ethnic groups; Catholics, Mormons, Baptists, and other religious groups; occupational groups; and so on. The second is learning the criteria by which people are classified as members or nonmembers of each of these groups. The third is the acquisition of appropriate modes of behavioral responses and attitudes toward each of these types of people.

*Howard J. Erlich, *Social Psychology of Prejudice* (New York: Wiley, 1973).

The criteria used by children to identify differences or reasons for differences change with age. Younger children tend to focus ethnic attitudes, for instance, on physical features such as skin color, clothing, language, and social customs. By age ten the reasons given for perceiving differences among groups shift to less observable criteria. Older children cite personality characteristics and ideological differences in religion and politics as distinguishing features of groups.

Children who are targets of negative stereotyping and its related expectations may reflect these negative expectations in their self-concepts. This can have adverse consequences since self-esteem has been related to levels of academic achievement, happiness, creativity, and the ability to form close personal relationships. There is evidence to indicate that "black" children acquire a negative image of black at an early age. A number of studies have found that a high proportion of black children, as well as of white children, assign positive characteristics to white rather than to black children. They also express play preferences for white rather than for black playmates in otherwise neutral test situations.

The sources from which children learn stereotypes are varied. Family attitudes are important, although there may be little direct instruction in attitudes and behavior. Television, books, movies, and peer groups—indeed, all the input from the society around them—add to children's categorization and stereotyping of groups of people.

Stereotypes are often difficult to overcome. Communications studies indicate that people interpret information to fit into previously held ideas. Given mixed information, people will only "hear" the part that supports the stereotypes they already have. Actually changing attitudes and behavior requires much more than the simple provision of factual material—people sometimes even believe they have no prejudices, but they still act in a prejudicial manner. Contact in nonstereotypical situations (such as finding that the local librarian moonlights as a belly dancer) will reduce stereotyping to some degree. Often, however, stereotypes concerning a group remain intact in spite of individual exceptions.

Review Questions

1. Give two examples of characters illustrating typical sex-role stereotypes in current television programs.

2. Why might it be difficult to change an individual's negative opinions about a group of people?

3. Which of the following sex-linked characteristics, typically ascribed to men and women in our society, have at least some biological basis according to the material in this unit?
    a. Ability to run fast
    b. Ability in math
    c. Aggression
    d. Ability to have children
    e. Ability to do detailed work requiring great patience
    f. Intellectual ability
    Can you assign the characteristics to men and women according to typical stereotyping beliefs?

4. How might a father treat his son differently from his daughter?

5. How might a prejudiced white twelve-year-old's description of a black child be different from that given by a white six-year-old child?

Questions Pondered by Psychologists, Sociologists, and Philosophers

Have traditional sex-roles been made obsolete by social and technological changes, such as birth control and a decreasing need for physical strength in the world of work and war? Is prejudice against one group or another universal?

Of Policy Matters and Public Interest

Will school integration reduce prejudice based on negative stereotyping? Those who favor integration say that stereotypes cannot be changed if there is no contact among groups to counteract them. The opportunity to make friends from different ethnic groups and to observe common elements of behavior in a situation such as a school enhances the possibility of real communication. They add that some of the stereotyping springs from the fact that some children are denied equal opportunity and that society should correct this. Opponents argue that because of their parents' race or income levels, children still tend to form friendships along group lines and that competition among groups may increase hostility. They add that bussing can increase negative attitudes because it brings in alien (nonneighborhood) groups of children who often come from a different socioeconomic background. They claim that all of these differences are accentuated through proximity.

Sex-Role Stereotyping

Sex-roles present a special case in stereotyping. This stereotyping is the most pervasive for it includes everyone. It is, to date, the most difficult to separate from biological foundations. The basis of learning sex roles may, according to the theory accepted, differ from other kinds of stereotype acquisition.

Sex-role identification is achieved at an early age. By age four most little boys and girls have fairly firm ideas of their gender and what is expected of them. This may occur through identification with the same-sex parent, the development of cognitive processes, imitation, or from basic biological processes. (*A Child's World* contains fuller explanations of each point of view on pages 349 to 350.) There is little question but that some of this identification does arise from innate differences between male and female humans, but how much and how it does or should affect behavior patterns is the subject of a great deal of controversy.

SOCIAL STEREOTYPING

Many studies of boys and girls have measured activity levels, sensitivity to sound, light, touch, and so on. Few clear and consistent differences have been found. Boys tend to be more physically active than girls. Girls are not, as is generally supposed, more sensitive to sounds, and boys are not more responsive to visual stimulation.

There may be a small tendency for girls to acquire language skills sooner and to excel in verbal ability, but the research findings are often unreliable. According to Maccoby and Jacklin, the authors of a major study of sex-role stereotyping, few differences can be found between the sexes in intellectual performance during the early school years. Boys and girls score equally well on a large variety of tests: creativity, analytical thought, concept mastery in Piaget's tasks, and moral judgments. Boys tend to show greater variability in school performance and tested ability—that is, more score high but more also score low. This may be a result of boys' greater vulnerability to the difficulties of growth—before, during, and after birth. Boys suffer more often from bed wetting and infantile autism. They exhibit more learning difficulties involving mental retardation, stuttering, and reading difficulties. Girls usually get higher grades in school than boys, but this could be a result of a number of factors other than sex.

Girls do not display greater interests in friendships or social activities than boys, but they often tend to be more helpful or nurturing. On the other hand, aggressiveness and competitiveness are more characteristic of boys during the early and middle school years. Of all the sex differences we have discussed, these last two are the most consistent with social stereotypes. Aggressive behavior and the tendency to dominate is also closely linked to high hormonal (testosterone) levels among young primates and rats, however, so there may be a biological basis as well as a social basis for such behavior.

Both *A Child's World* and the television program provide examples of ways in which children are socialized to sex roles and of the resulting stereotypes. Boys are encouraged to be masculine, whereas girls are rewarded for being feminine.

The prohibition for tomboyishness among girls is much less than for girlishness among boys. Psychologists have indicated that the masculine role is much more restricted than the feminine in current American society; fewer explanations are required, for instance, if a woman works than if a man stays home and takes care of the house.

The negative impact of such role stereotyping may be the restriction it imposes upon a man or woman's choice of behavior. In addition, many women suffer from negative self-esteem as a result of some people's attitudes toward their capabilities and society's low valuation of certain "feminine" character traits. Men, on the other hand, may suffer if they do not meet society's expectations of "success" or if they cannot provide for a wife and family. A firm sense of sexual identity, however, is associated in many cases with positive self-esteem. Just as the question of whether sex differences are innate or acquired is still open, the question of how much and what kind of sex typing is beneficial is yet to be answered. Ultimately the answer may depend on the context provided by society itself.

17

Play

Student objectives

1. Discuss the main theoretical interpretations of the importance of play.

2. Describe the functions of play as they relate to physical, mental, and personality development.

3. Describe the types of play characteristic at each stage of development (ages).

4. Discuss imaginative play or fantasy as an expressive mechanism for children.

Assignments for this unit

1. Read pages 322–329 in *A Child's World*.

2. Watch Program 17, "The Child's Play."

3. Read Overview 17 in this book; this overview is designed to extend the rather short text assignment.

4. Review the study aids.

In the television program Robert Goldenson defines play as "any activity that is freely chosen and pursued for the sake of sheer enjoyment." Note the choice of words. Play is an activity. It is freely chosen rather than something demanded of children. It is "pursued for the sake of sheer enjoyment." How does an activity as light hearted and *fun* as play become the subject of scientific study?

The impetus, perhaps, began when the question "Why do children (and adults) play?" was first asked. What purpose does play fulfill? Where does this really quite complex and mysterious activity fit into human development?

Theories of Play

Playful behavior is certainly not confined to humans. Kittens and puppies spend much energy in play; detailed studies have been made of the play activities of monkeys, chimpanzees, and other primates. The content of animals' and children's play originally gave rise to the evolutionist explanation most commonly associated with Karl Groos. According to Groos, who published his theories in 1896, play— the kitten pouncing on a ball of yarn, a chimpanzee poking a stick in a hole, or a child pretending to hunt a deer—was nature's way of providing preparation and practice for the activities of adult life. As such, play is an instinctive form of behavior. The amount of play needed depends on the position of the species in the animal kingdom. Kittens, being higher than ants, need play much more than the insects. Children, in the same manner, need more than kittens. Groos related this need to the complexity the animal (or human) could expect to find in adult duties and pursuits.

Even before Groos, Herbert Spencer had propounded the surplus-energy theory of play. A child or animal has energy that must be released in some manner. Since useful or practical skills have not yet been learned, the energy is expended in play. Another nineteenth century figure, G. Stanley Hall, suggested that play recapitulated the history of the

ne notice is given of the other person through actions and
rds.

4. Associative play. Children interact and cooperate to
ne extent. They share materials and their conversation
icerns a common activity, but each child is concerned
marily with his or her role rather than with the functioning
the whole group.

5. Cooperative play. This is group play in an organized
ivity. The play group itself is usually goal oriented with
ferent roles, some of which are subordinate, taken by each
d.

Most preschool children engage in all types of play, but
re seems to be a positive correlation of the amount of play
each category with the child's age. Younger children en-
ge in more solitary play, older children in relatively more
perative play. There is, however, some ambiguity as to
ether progression from one type of play to another occurs
he same rate for all children. The two studies cited in the
t (Parten and Barnes) found differences in the number of
he-age children at each level. Other studies have found
erences in the solitary or social play preferences of pre-
ool children according to both sex and social class. The
hors of still another recent article on these categories of
y found that over 50 percent of preschoolers' solitary play
educative or goal directed in nature. According to earlier
lies, solitary play is associated with younger children and
ne of the least mature forms of social play. This last study
stions whether this should be so; it may be that parallel
y rather than solitary play should occupy this position.

Classification of play on social grounds includes some,
not much, indication of the content of play. One of the
e recent classifications of play identifies types of play in
ns of content. Within each category the study also follows
nges in the way the content is approached by children as
y grow older. The categories, according to Brian Sutton
th, are:

1. Imitative play. This begins in infancy when the par-
and baby imitate each other. As children grow older,

human race. When children pretended to hunt and farm, they
were passing through the stages their ancestors had lived.
According to Hall, a human must live through each stage
before he or she could progress to the next. These stages, of
course, culminated in responsible modern adulthood.

Sigmund Freud represents, perhaps, the first "modern" theory of play. According to the Freudian psychoanalytic
model, play serves a number of functions:

1. Play is a source of gratification. Children express
the pleasure principle through their play. They tend to seek
pleasure, and play is fun.

2. Play also acts as a cathartic experience. If a child is
disturbed by the emotional impact of certain experiences, it is
possible to "play" at those experiences until they can be accepted and assimilated. The *cathexis* of objects—or the in-
vesting of psychic energy (libido) into an object—indicates
that the toy or object has some special significance. In these
terms, Linus' blanket could be a symbol of, and substitute
for, his mother's breast.

3. Erik Erikson added a third important element to the
psychoanalytic view of play. As reality begins to intrude on
the child's world, play helps in the conflict between id and
superego. Play, in helping the child reconcile her drive to
seek pleasure with the demands of reality, serves to build ego
strength. Erikson also brought the element of play as preparation for adult activities into Freudian theory.

Young children's play is particularly interesting to
Freudians because the symbolic use of actions and words
reveals conscious and unconscious thoughts. Furthermore,
the unconstrained act of self-expression in an unthreatening
situation helps the young ego to acquire independence from
the id.

Jean Piaget represents yet another approach to play.
For him play is a means of assimilating new information into
the child's world. The type of play is dependent on the
child's stage of cognitive development. Piaget identifies three
major types of play:

1. *Practice play* appears during the sensorimotor stage and includes actions that are ends in themselves. These include such things as climbing stairs, rolling over, and, later, running and jumping.

2. *Symbolic games* involve a more sophisticated level of thought than practice play and appear slightly later. They can be individual or social and imply a comparison between reality and an imagined element. They can also imply make-believe representations.

3. *Games with rules*, the next level of play, imply social relationships with rules that are imposed by the group. These games generally appear after the child has entered the stage of concrete operations.

Piaget, like Freud, believes in the importance of symbols to the child, but he largely ignores the distinction between conscious and unconscious thought. Piagetian symbols aid the child in forming mental representations of external objects and events. Objects—three blocks—serve as symbols of what the child wants to think about—an engine, a box car, and a caboose. In Freudian theory, on the other hand, the use of symbolism in play permits expression of conflicts and feelings of which the child has little conscious knowledge. The three-block train in this example could be a means of dealing with Mama's departure by train and the child's feeling of loss.

Play, in Piaget's theory, is adaptive. Repetition, a common feature of infant behavior, clearly serves to strengthen sensorimotor patterns of thought (schemata) during the sensorimotor period. Novelty, a source of great pleasure to most children, helps the child discover new facts, perceptions, and sensations. Both reflect Piaget's feeling that play, like the other activities of children, is biologically useful. Play is an exercise of a function or skill; it is a way to learn about new and complex objects and events, to consolidate and enlarge concepts and skills, and to integrate thinking with actions.

Learning theory holds that play is learned behavior, is reinforced by adult approval, and often reflects the cultural values of adults who influence children.

Theories concerning the functi[on]
are many variations on the major
mentioned—are sometimes contradicto[ry]
or they contain similar elements. Som[e]
from differences in the definition of pl[ay]
differences in the approach to wider
such as personality or cognitive devel[opment]
theme, however, that runs through m[ost]
gists' characterizations of play: play is

Types of Play

One way that has been used to
of play is to observe *how* children pla[y]
play behavior, then attempt to relate t[o]
of development. One of the results ha[s]
of types of play.

There are two major classific[ations]
terms of the social character of play,
with and how the children interact.
as for Piaget, with the stages of intel[lectual]
more simply in some other psychologi[cal]
second classification is of the conte[nt]
child is doing.

A synthesis of the classificati[on]
the following stages:

1. Imitative play with adults.
as bye-bye or making faces in respo[nse]

2. Solitary play. Children p[lay with]
their own toys and make no effort
nearby.

3. Parallel play. Children p[lay and]
their activities bring them into conta[ct]
children at this stage are still basica[lly]
ter, they usually play alongside eac[h]
teraction but, as in *collective monol*[ogue]

they may pretend to go to the office like Mommy or drive a car like Daddy. Gradually this merges into dramatic play where adult roles—cowboy and Indian, schoolteacher and pupil—are imitated.

2. Exploratory play. Infants begin exploring their environment at an early age. Preschool exploratory play may involve trying out familiar objects—sand, mud, clay, words—in new situations as children explore novel feelings, effects, and relationships in a familiar setting.

3. Testing play. In many types of play children test themselves. Physical activities are often among these—running faster, jumping from higher walls, throwing a ball alone or in competition with a playmate. Memory games and games of choice illustrate different areas where testing can occur through play.

4. Model building. Creating representations of the real world in a combination of imagination and experience occurs when children play with model worlds such as those created by a train set. Models are not only physical but often include elaborate schemes of social relationships, rules, and events.

These categories of play are descriptive in that they do not relate to a theory of development. The category of *imitative* play, for example, includes imitation of mother's facial expression in the sensorimotor period as well as the more deliberate and complicated imitation of social *roles* and people which occurs much later. Whether these examples of "imitation" are at all related is yet unknown.

In general, it seems that *categories* of play illuminate the varied uses of the concept "play," but unless they have theoretical significance in a developmental framework, they do not provide guidelines to the child's activities as reflections of underlying developmental processes. They do, however, provide a means of representing the different ways in which children explore and attempt to master their environment. Reflected in the categories, in the other classification schemes, and in the various theories of why play occurs, are some reasonably straightforward, easily identified effects of play in the areas of physical, mental, and social development.

Play helps children learn about their bodies and how they react to their surroundings. Active play develops muscles and motor skills. Exercise—in this case, gained through play—is necessary for good health and physical development.

Children learn more than motor skills through play. They develop concepts of size, shapes, colors, textures, mass, and time. They learn to classify items or to put them in sequence. They practice all these skills in different settings and different combinations through play at various games and with a variety of objects. Through play children can create a world, albeit make believe, and increase their understanding of the real world.

As children begin to play with other children, they develop social skills. Peer groups bring pressure that is often more effective than that of adults in establishing codes of conduct and rules of behavior. Along with social skills, play aids the development of moral standards. "It's Jennie's turn," "That's not fair!" "Brian's a cheat," echo social concepts of morality reflected and reinforced by the play society.

Study aids

Review Questions

1. In which of Parten's categories would the following types of activities fall?
   a. A lone child building a tower of blocks
   b. A game of hide and seek
   c. "Playing dress up" where each of three children concentrates on finding his or her own clothes

2. In which of Sutton-Smith's categories would these types of activities fall? Do his descriptions of these categories tell you anything about the age of the children in each case?

3. Debbie is playing doctor and giving her doll a shot just like the one Debbie had the day before. What would

this activity signify according to the psychoanalytic theory of play?

4. How would the rules for tag set up by a group of children aid personality development?

5. In Piagetian theory, what kind of games would a child in the sensorimotor stage play?

6. Identify two or more functions of a fantasy game in which children pretend they are bus driver, policeman, and passenger without a ticket.

7. How did the groups of children in Barnes' study differ from those in Parten's? To what did Barnes attribute these differences?

Questions Pondered by Psychologists and Educators

When does play stop being play? Is collecting stamps play? Practicing setups for the volleyball team? What is the relationship between a child's play and work?

Of Policy Matters and Public Interest

How much time do children need to play? As children grow older, increasing amounts of time must be spent learning the skills—at the most basic level, reading, writing, and simple computation—that will allow them to operate in adult society. Often doing well in school requires homework. Students join clubs and participate in sports. Some time must be spent in doing chores around the house. At each age there is a question of the proportion of time needed for each type of activity—learning, responsibilities to the family, organized activities, and play.

Importance of Fantasy and Expressive Play

Fantasy is closely related to many forms of play. The girl pretending she is a cat, the boy building a bridge with blocks, and the children pretending to play restaurant are engaged in both fantasy and play. The two, however, cannot be equated: The shortstop in a ball game or the boy building a row of dominoes and knocking them down to see them fall are not engaging in fantasy. The harried wage earner dreaming of winning a lottery is not playing.

The importance of fantasy to play is great. Without fantasy or pretend, play would be impoverished in content. It would also be impoverished in function, for although the precise role of fantasy in play is not clear, it seems to be important in many ways. Most of these relate, of course, to the child, but children's fantasy play can also be used by adults, either to learn about children or to learn about themselves.

Rather than tie the possible role of fantasy and expressive play to theoretical approaches, we will outline areas in which it seems to have particular value. These areas have been identified by a number of psychologists. Few would characterize fantasy as having only one function, but the theoretical approach of each might lead an individual to emphasize one aspect rather than another.

The Freudian approach is that fantasy has a cathartic effect on children. Through this play children are able to express and deal with fears, aggression, and other emotions. Imagination is not necessarily a way to avoid problems but can be a means of overcoming them.

In a related, but not completely equivalent, function fantasy play allows children to engage in behavior that would otherwise be unacceptable. "Bang, bang, you're dead," "I'm Daddy and I can say 'Damn' ", or "I'm a horse and that's why I'm eating grass," may be ways for children to do things that they would not be allowed to if simply "themselves."

Through dramatic play and fantasy children try on adult roles and play at their life's work—growing up. Imaginary playmates provide companionship and, often, can be endowed with attributes children would like to have for themselves.

Fantasy and expressive play may be important to the development of creativity and imagination. The formation of images practiced through fantasy play may aid the child to store and comprehend complex concepts.

Finally, fantasy and expressive play may provide adults with an indicator of the children's feelings and thoughts. Personality and adjustment problems in young children can be uncovered in psychotherapy by analyzing play activities for significant fantasies.

18

The Preschool Experience

Student objectives

1. Describe various types of preschool programs for young children.

2. Discuss the basic needs of all children in preschool programs.

3. Discuss the effect of the working or professional mother on the growth and development of the preschool child.

4. Describe some possible values and negative effects of a nursery school experience.

Assignments for this unit

1. Read pages 285–296 and preview pages 523–526 in *A Child's World*.

2. View Program 18, "The Preschool Experience."

3. Read Overview 18 in this book and review the study aids.

Sally's mother took ten days leave from her work when Sally was born. Since that time, with the exception of vacations and periods when they were changing jobs, both Sally's parents have worked full time. For the first 15 months of the baby's life, the family lived abroad with a full-time nursemaid and Sally's mother's office in their residence. When they returned to the United States, Sally spent weekdays with a nearby family, acquiring a "brother" her own age in the process. Now that she is three and a half and the family has moved again, Sally trots happily off to her "office" every day—a nearby preschool—"just like Mama."

The concerns of Sally's parents throughout these various child-care situations reflect the concerns of many parents whether they work or whether they hope to enrich their children's lives through appropriate use of a nursery or preschool situation. Sally's parents want her to develop in all the ways that we have seen are important to infants and children of this age: to be able to explore the world around her from a base of warmth and security; to form good relationships with adults and with other children; to grow intellectually, emotionally, and physically in a manner that is appropriate for her age.

These goals are not unusual. Although they may word such goals differently, most parents have them for their children. Questions that arise are usually approached and answered unconsciously in most homes. In the nursery schools and, increasingly, in day-care situations, the questions are asked and answered on the basis of educational and psychological theory.

History

Five or six has traditionally been the age to begin schooling. Long before Erikson or Piaget outlined their stages of cognitive and emotional growth, philosophers and psychologists recognized the need for an approach to learning for the young child that was different from that used at later ages. Plato specified the kinds of stories, dramatic poems, and music that would build character in preschool children in ancient Greece. The "School of the Mother's Knee" of John Amos Comenius provided seventeenth century preschoolers with training in the use of the senses and a mastery of words and elementary facts. Jean-Jacques Rousseau and J. H. Pestalozzi both wrote that children should be allowed to develop naturally, but the idea of a special school for children under six years of age was not systematized with a theory behind it until Frederick Froebel (1782–1852).

Froebel originated and named the "kindergarten" or child's garden. Behind the various games and activities that children pursued in his school was Froebel's belief that the child's potential would unfold through self-development. This approach has remained a major influence in preschool and primary education until the present. The preschool years are seen as the time to play and to learn to play with other children. One educator suggests that the objectives of kindergarten education can be formulated as follows:

Friendliness and helpfulness in relationships with other children.

Greater power to solve problems based in individual activities and group relationships.

Respect for the rights, property, and contributions of other children.

Responsiveness to intellectual challenge.

Achievement of good sensorimotor coordination.

Understanding of concepts necessary for the continued pursuit of learning.

Responsiveness to beauty in all forms.

Realization of individuality and creative propensities.*

*Neith Headley, *The Kindergarten: Its Place in the Program of Education* (New York: Center for Applied Research in Education, 1965).

A recent trend has focused on the belief that younger children can and should engage in more structured "learning" of a cognitive nature. Currently there are educators on either side of the cognitive/affective spectrum, but both groups are, perhaps, being superseded by theorists who question whether the two have to be separated. According to this last group, young children grow and learn in all areas: cognitive, emotional, social, and physical. The kindergarten or, for younger children, the nursery school and day-care center should facilitate all-around development.

Theory

This resolution, however, often leads to a more lasting controversy concerning education, one that continues throughout much of the child's and adult's education. In discussing this controversy, we shall identify two general approaches to the question of learning and link them with bodies of psychological theory we have been studying, but you should remember that there are many variations and combinations of these themes.

The first dichotomy concerns *product* versus *process*. If, by the end of the first year of preschool, a child should be able to count to ten, recognize colors, and tie her shoes, we are looking at a *product* brought about by this first year of learning. The child has attained a certain, specifiable level of achievement or competency. In the *process* view, education is a life-long affair and everything contributes to it. The child who has learned to count to ten may not have learned as much as a child who can only count to five but who has gained much more in other areas.

There are problems in both approaches, and the two need not be mutually exclusive. Product-oriented education may be confining—identifying some components of a "good" educational experience may leave out too many other factors and limit the student in ways not seen by the teacher. A process-oriented educational approach may, on the other hand, be terribly vague. If students and teachers have no guidelines and goals, they can wander through the world of things that conceivably could be "learned," without accomplishing or gaining as much as they want to. Indeed, they may not know what they want to accomplish or gain. In many cases educators attempt to combine the two. The student objectives at the beginning of this chapter are product-oriented; they specify items you should know or be able to do at the end of the unit. A number of references in the study aids are process-oriented; they have been designed to ask you to think about what you have learned in terms of your own life and other material, to relate these facts to each other, and to realize that they have more lasting value than as merely an aid to good grades and college credit.

Product-oriented education is, to a degree, associated with the behaviorist approach to learning. Process-oriented education is often linked to those who feel that the child learns as a result of both maturational processes and interaction with the environment.

Behaviorists believe that learning derives from the responses of an individual to outside stimuli. This is a type of approach in which age is not particularly important; the type of learning that can be achieved depends on what has been learned before. It is also an approach in which objectives must be established; in order to modify behavior the teacher must know what outcomes or resulting behavior the child should acquire. Once the behavioral outcomes are established and the student's current level of behavior is assessed, the teacher provides the child with the proper types of stimuli—toys, games, instruction—to achieve the desired result.

At the other end of the scale, maturationists believe that children's ability to learn is dependent upon maturational processes; learning of some concepts and activities cannot take place until the child is ready for them. Although not strictly a maturationist, Piaget is one representative of this type of approach. In Piagetian theory the environment provided the child is important for learning, but it must be geared to the child's degree of mental maturation. This approach does not preclude the formulation of specific learning objectives, but it is less often associated with them than is the behaviorist approach.

Compensatory Preschool Education

We have seen that infants and preschool children from materially or emotionally impoverished environments tend to do less well in general development than children from more secure backgrounds. In other units—particulary Units 21 and 24—we shall see that these chilren not only start school with a lower achievement level but that the gap between them and most other children normally continues to widen as the years go by.

A number of psychologists and educators have suggested that compensatory education during the preschool years might be a means of remedying the handicap these children have. Operation Head Start, begun in the early 1960s, was a federally funded attempt to do just this. To a large extent, programs were locally controlled, but they shared one major common denominator: their students were children whose socioeconomic status predicted marginal success or outright failure. Through Head Start programs these children were given the opportunity to learn many things other children learned in their homes or in regular preschool programs—for example, counting and simple skills such as cutting with scissors or holding a pencil. In addition, they were often given general enrichment experiences (toys and games, outings) and health care.

The first major reports on the results of the project were mixed. According to such studies as the Westinghouse Learning Corporation Report, children generally showed gains during the program but failed to retain them once they entered regular school. Explanations for this ranged from criticisms of the studies to blaming the regular schools for not continuing Head Start's innovative, effective approach. Some psychologists felt that Operation Head Start, aimed primarily at four- and five-year-olds, did not begin early enough to undo the effects of environmental deprivation that had already been established.

More recent studies of Operation Head Start and of Operation Followthrough—a program begun in the late 1960s in an attempt to help children retain gains under Head Start once they started school—have been more positive. Additionally, psychologists have found a number of side effects emerging from the Head Start studies. Before 1965 we knew very little about poor children and less about how to evaluate them and project what extra help they needed. Head Start data have now added greatly to our ability to diagnose problems and prescribe appropriate activities. Head Start has also affected the attitude of the general public. People are now much more knowledgeable about the handicaps of some children and are concerned about remedying them through compensatory education.

Problems in Preschools

Preschools and day-care arrangements can be good and bad. Generally, the criteria outlined in the television program are basic to any kind of care for children, inside or outside the home: a concern for the child's health and safety, an adequate number of adults who provide care, and an environment that assists the child in positive mental and physical development. Beyond that, most parents must decide the kinds of things they wish to look for. Many turn to psychologists, who have touched on a number of areas in which the use of day-care or preschool facilities can be of concern.

A preschool must be congruent with the child's home situation. This often affects children from lower socioeconomic backgrounds, but it can affect others. If the values, attitudes, and behavior required at a preschool are completely different from those required at home, children may have problems relating to the two environments and relating the two environments to each other.

Spending a large part of their waking hours with other children in a preschool environment may limit children's horizons rather than broaden them. Contact with adults is reduced, and the peer group becomes important at an earlier age than usual. Children's views of adult activities outside the preschool are circumscribed because the children are not in contact with adults for most of the day.

Parents may come to feel less responsible for their children's welfare and behavior than they might if the children were at home all day. There has been a tendency to ask schools to "do everything" for children, but many educators feel that much early learning is best provided by the home. This is particularly true in the areas of moral development, general attitudes, and behavior. Some psychologists feel that having a child in preschool may lead inexperienced parents or parents whose attention is engaged elsewhere to neglect their own educational function with their children.

Some children may suffer when separated from their mothers. However, we do not know exactly what effect the kind of separation engendered by preschools has. Most of the studies of separation anxiety and maternal attachment have dealt with children who were completely separated for long periods of time. There does seem to be evidence that such separation should begin before six months of age or after about eighteen months (psychologists differ greatly on the last figure) because children are usually particularly anxious about strangers during this period. Since this is earlier than the age at which most preschools in the United States will accept children (normally they insist that children be two and a half and/or toilet trained), it does not affect the preschool child, but the question remains an important one for working mothers of very young children.* There is also evidence that short periods of separation are beneficial to many preschool

*An area in which more research is badly needed concerns the role of adults other than parents. We do not know how another family member's care (such as a resident grandmother or a father) differs from a mother's or how a sympathetic neighbor's steady care might differ from a family member's in its effects on the young child.

children and that a good preschool program enhances their all-around development.

Study aids

Review Questions

1. One preschool rewards children for good behavior with tokens which can "buy" treats and previleges. Another advertises that it doesn't believe in teaching reading because children are not yet "ready." With what psychological theories of mental development might each be associated?

2. Contrast each of the following kinds of preschool programs:
   a. Nursery school
   b. Day care in a licensed home
   c. Kindergarten
   d. Head Start

3. Sally's parents felt she should spend her days in a family even if it were not her own rather than a preschool before she was three. What factors might have influenced this decision?

4. Who was Frederick Froebel? What relation did he have to Rousseau?

5. Contrast the *product* and *process* approaches to education.

Questions Pondered by Psychologists and Educators

When is the best time to begin education for children from underprivileged environments? Should very young children be taken out of the family when family situations are bad?

Of Policy Matters and Public Interest

What should parents look for in a preschool? Some items are standard. Parents should ask about toys and equipment for variety and safety; they should ascertain the student–teacher ratio and make sure that the child is given nutritious snacks or meals in clean surroundings. The personality of the individual teacher is very important. Parents should ask about means of discipline. Programs vary, but so do children. Some children respond well to relatively unstructured programs, whereas others are lost. Often choices—and good ones—are made simply on the basis of how parents "feel" about a particular school, its facilities, program, teachers, and convenience to places of work and home.

19

The Child's Mind— Part 1

Student objectives

1. Describe the characteristics of the child's thinking in this period, including Piaget's stage of concrete operations.

2. Describe the development of concepts of *time, space, number,* and *classification* in six- to twelve-year-old children.

Assignments for this unit

1. Review pages 273–285 and read pages 411–422 in *A Child's World*.

2. View Program 19, "The Child's Mind—Part 1."

3. Read Overview 19 in this book and review the study aids. Note that development of the concepts of classification and number are discussed on pages 281–283 of the text and in Overview 13.

4. Read Causality in this chapter (optional).

This is the first of two units on the mental development of children between the ages of six and twelve. In them we will spend most of our time discussing the theories of Jean Piaget not only because his is the most completely articulated theory of development, but also because it is more age-specific than most. Behaviorists and others describe many of the same phenomena as Piaget, but their basic approach to the character of thinking and learning covers all ages. Piaget, on the other hand, perceives children as moving through qualitatively different stages of mental development. His characterization of each stage and his description of the activities a child can perform in it make it necessary to discuss both the observable behavior of children and his approach to it in some detail.

Significant changes begin to occur in children's patterns of thought when they are five or six years of age. Between then and age eleven, children will be in the process of evolving true concepts out of preconcepts. In Piagetian terms, *animistic* beliefs dissipate as children learn that not everything in the world is alive and as they progressively draw the boundaries that distinguish living from nonliving beings. Children acquire *reversibility* of certain mental operations and can progressively *decenter* as they learn to consider all aspects of a situation rather than concentrating on one point to the exclusion of all others. By the end of this period of *concrete operations, formal thought* in Piagetian terms, will come to provide a means of transcending practicality by considering pure possibility.

School is the backdrop to the progression of cognitive development in the middle years, not only in the formal teaching situation but in the informal groups of children. Children have to learn to respect the rights of others and assert their own. Their adaptability is tested, judgments challenged, feelings bruised. Fantasy and idiosyncratic thought give way to a new awareness of shared communications and activities. With the clash of different points of view, the *egocentrism* of the preschooler is gradually overcome, though more slowly in some areas than in others.

In a primary class of 30 to 40 children, individual instruction is difficult and, since some children are more "ready" than others for the first lessons, a few always seem to be left behind. This may have less effect on the child's mental development than first appears, however, since evidence is accumulating that cognitive development up to about age seven or eight is not greatly affected by formal instruction. Children will learn by self-discovery if they are given the opportunity. Reading is taught formally in the United States around five or six; in some western European countries educators wait until a child is seven. Early reading training does not seem to influence the way children think, though some knowledge may be gained.

Much of Piaget's research on the changes in mental development that occur during these years has centered on *conservation* tasks. The knowledge of physical properties, taken for granted by most adults, is acquired over a long period of time by children. The course of this acquisition is fairly stable, and psychologists have found that studying it aids understanding of the mechanisms of thought.

One important conservation experiment involves two identical beakers filled with equal amounts of water. While the child watches, water from one is poured into another, empty, beaker, which is shorter and wider. The experimenter than asks if the new beaker has more water, less water, or the same amount of water as the first, unchanged beaker.

Typically, children under six or seven years of age will give an answer that depends on the shape of the new beaker—in this case, it is usually "less." Piaget claims that this illustrates *preoperational thought*. It is not *reversible*, for the child clearly does not mentally picture the water being poured back into the first beaker; it is still tied to the *static* pictorial view. The child also *centers* on one aspect of the experiment, most often the height of the water in the beaker.

Around age six or seven children enter an intermediate stage where inconsistent responses are given. By seven or

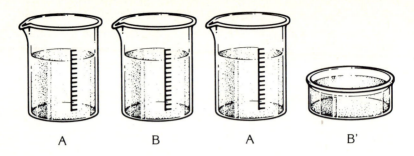

A   B   A   B'

eight most children acquire the concept of conservation of liquids, although other conservation tasks may be mastered earlier or later. In all of these, children's answers to questions about their responses demonstrate several properties of operational thought:

 1. "The water is the same since you can pour it back again to the same level." This argument represents *reversibility* or a *negation* operation of thought; the opposite action is seen as a means of restoring the original situation.

 2. "It's the same because none was added or lost." Piaget claims this argument is based on a mental operation that preserves *identity*. Since it is the same object, its properties must be the same.

 3. "The height is less but the width is greater, therefore the quantity remains the same." Reciprocity or *compensation* in thought implies that the child recognizes that a change in one dimension is compensated by an opposite change in another dimension.

 These qualities of thought are not necessarily transferable to other conservation tasks. As the text points out, conservation of quantity and weight occurs later than that of substance; this inability to apply mental operations used in one area to a different area is called *horizontal decalage*.

Piaget would describe these changes in children's thought by saying that the child is incorporating new mental structures, or *schemata*. Some criticism has been made of the vagueness of the idea *schema*. According to Piaget, schemata are sensorimotor or operational structures; a schema develops on the basis of actions and interactions with the world. This view seems plausible for understanding the perception of "space" (actually surfaces and objects) since actions are directed toward and by the physical environment. These principles are not always so satisfying in the explanation of the growing understanding of complex events, such as social situations or occurrences at a distance. Clearly not everything can be learned on the basis of sensorimotor or operational schema. Indeed, one psychologist, James Gibson, has developed a theory of space perception that denies actions are required, and much evidence has been presented to support this view. Piaget has countered this criticism to some extent in his examination of *event perception*. He approaches the comprehension of events through a complex analysis of the child's perceptions of *space* and *time*.

Piaget feels that time is grasped operationally when several related concepts are achieved: simultaneity, duration, and distance. He found, for instance, that children younger than five or six could not arrange drawings of water being poured from one flask to another in sequence even though they had watched the experimenter do the pouring. His explanation for this was that they could not relate the series in time. "This was first, then this, then this, and finally that." Their reasoning was based on a conception of the event that was comprised of static images.

In another example, if one object moves faster than another but both start and stop at the same time, younger children will say that the faster-moving object was in motion longer. Children center on speed and do not relate speed with time and distance simultaneously. Correct judgments of these features occur at about the same time, indicating that a coordinated schema of time has developed.

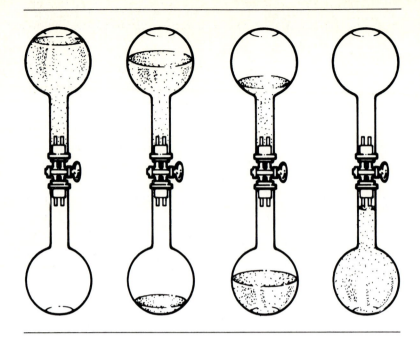

Other experiments deal with subjective or *psychological time*. Most of us have felt that time flew or dragged on one occasion or another. Learning to step back and make objective judgements of the minutes or hours that actually passed while doing something requires *decentering* from one aspect of the task and detachment from one's own feelings about it. In asking young children to judge how "long" they spent in doing certain tasks, Piaget found that younger subjects tended to base answers on the speed at which they worked or on the final results. "I put lots of matches in this pile and it took me a long, long time."

Another aspect of event perception is the child's growing knowledge of space. Preoperational children think in "topographic geometry" rather than "Euclidean geometry." They recognize the existence of concepts such as inside–outside, near–far, line–closed figure, and so on, but do not seem to grasp concepts such as straight line, horizontal and vertical axes, angle, or curve. It is not simply a matter of perception. Young children can, for instance, tell the difference between a circle and a square, although they cannot reproduce a square.

They also have difficulty coordinating various aspects of a figure against a background. If shown a jar with colored water, they do not realize that the water level remains horizontal even if the jar is tilted. Very young children will not even draw a water line, although they will indicate that the water—represented by scribbles—is inside the jar. Older children will draw the line, but it will be parallel to the bottom of the tilted jar, not to the table.

These specific conceptual operations relate to changes in children's thinking in this period. Other, more general, properties of thought are also developing in the middle years. Some of these will be discussed in Unit 20, which should be treated as a coordinated unit with this one.

Study aids

Review Questions

1. Assuming that the stimulus was desire for the experimenter's approval, how would behaviorists explain a child's learning to *conserve* water? How does this differ from Piaget's explanation?

2. How does a child's conception of space change from Piaget's preoperational stage to the stage of concrete operations?

3. "Dolly doesn't want to open the refrigerator door. She thinks the freezer wants to eat her up." This statement is an example of what characteristic of preoperational thought?

4. Does a child master all aspects of operational thought at the same age? Explain your answer.

5. If A is greater than C and C is greater than B, what relation does A have to B? By what technical term

would you describe your ability to give the correct response?

6.   The stage of concrete operations is characterized by:
   a. The ability to decenter
   b. Increased logical thought
   c. The understanding of reversibility
   d. The appearance of mental operations
   e. All of the above
   Can you illustrate the correct answer by examples?

Questions Pondered by Psychologists

When confronted with six roses and four tulips children will insist that there are more roses than there are flowers. Is Piaget's explanation for this phenomenon adequate? Do they mishear over and over again? Have they been taught not to double count so thoroughly that they mistake this for double counting?

Of Policy Matters and Public Interest

These are the years when the psychological theories held by educators bear directly on children. Teaching reading and language skills is one of the most widely explored—and perhaps the least understood—examples. Drill, repetition followed by reinforcement, is characteristic of the teaching approaches based on behaviorist philosophy. Other theorists suggest teaching must be tied to student interest and built on student activity and involvement. Both approaches work with many children, whereas other children respond to one or another. We are not quite sure why this is so.

Causality

The search for cause-and-effect relations is common among young children. Innumerable and persistent *why* questions widen their knowledge of the world. Contingent events—events that tend to occur together in a particular order—and consistent results of particular actions are noted and learned.

To give an oversimplified example, we can say that causality can be mechanistic (the vase broke because I hit it with a baseball bat) or purposive (I hit it because I couldn't stand the color). Children begin an analysis of causality by ascribing a purpose to all events, even mechanistic ones. The clouds move because "they want to go over there." The physical world is seen as an extension of the child's own experience, and this leads to one of the earliest explanations of causation. Events occur because of *animism*—the objects involved in them are alive.

In young children up to about six or seven, everything that functions in some way is thought to be alive. Later, only moving things are regarded as animate. At about ten to twelve years of age, mechanistic events are separated from the animate, but even then events that do not have a perceptible cause are considered alive, as in the examples in the text.

Artificialism is related to children's interpretations of their world. This refers to the tendency of egocentric children to attribute final cause to themselves, their parents, or to people in general. Only at around nine years of age are children able to understand that people did not create the world.

Piaget sees the development of true concepts of causality as a result of developing cognitive schemata. Another theorist, Michotte, developed an experiment where one block seemed to hit another, which then started to move. As a result of responses to various sequences of these events, Michotte decided that perception of causality is dependent on the inborn organization of time and space. The Gibsons (mentioned in the optional section of Unit 13) believe that changing reasons for causality in this type of example are based in increasingly sophisticated perceptions—children come to perceive that one action produces another. Short-term and long-term events are eventually perceived through a process of discrimination. The Gibsons think that the notion of *causality* is a term we apply to our experience of certain special events and that it should not refer to a mental construct.

20

The Child's Mind—Part 2

Student objectives

1. Discuss language development in the middle school years.

2. Describe the developmental processes related to reading and identify some of the sources of common reading problems.

3. Describe the development of memory and memory skills in this period.

4. Describe creativity and identify its components.

Assignments for this unit

1. Read pages 437–440 and 448–476 in *A Child's World*.

2. View Program 20, "The Child's Mind—Part 2."

3. Read Overview 20 in this book and review the study aids. Pay particular attention to the section on language development in the text as it is not repeated in the study guide.

Many of the aspects of mental development we are studying in the middle years section of this course are not unique to the ages between seven and twelve. No one would argue that creativity, memory, the ability to read, or the use of language are special characteristics of this period. Nevertheless, during the middle years all three develop in ways that will be important to the child's future, and this seemed a good point at which to examine them.

Creativity

Creativity is an elusive concept. We can describe some of the characteristics of creative thought or of those we term creative individuals, but we cannot draw a boundary, saying "This is creative" and "This is not." Indeed, the ability to leap conventional boundaries may be one of the special aspects of this form of human activity.

A creative person is one who possesses the capacity for new ideas. The thinking processes of a creative individual illustrate flexibility in finding new and unusual solutions to problems. Creativity also suggests productivity—we assume that the new ideas and approaches formulated by the creative person will serve some purpose. A schizophrenic in a mental hospital may generate many novel ideas, but these generally fail to have much impact on others and will often seem meaningless.

Theories concerning the source and development of creative thought are not precise nor are they capable of verification as yet. With his theory of the unconscious, Freud provided some of the most fruitful insights into the mechanisms of creativity. He believed that conscious thought was often directed through unconscious processes. Artists, writers, and others with a rich, imaginative life are thought to be more attuned to their unconscious. Rational thought, on the other hand, is constrained by logic and conforms in greater measure to the demands of the real world. This secondary process of thought, as Freud called rational thinking, develops later in children and overlies the primary process of early egocentric thought. It may be the developmental process itself that explains why children seem to reach a peak in creative imagination in about the fourth or fifth grade, whereas only a few adults retain these characteristics to any extent.

Memory

Memory as well as creativity develops during this period of childhood, not only in what children remember but in their ability to organize information into memorable structures. In the most basic sense, memory refers to the ability, common to all animals, to alter behavior as a result of previous interaction with the world. Memory is inseparable from learning; yet memory and learning are separated by contemporary theory. In modern cognitive psychology the term *memory* refers to acquisition, retention, and organization of information for storage. Behaviorists tend to reduce the memory function to the "learning of associations" between stimuli and responses.

Psychologists have traditionally studied learning and memory in a number of settings. Each of these provides some insight into the nature of memory and to the problems psychologists have had in defining and studying it.

1. *Pavlovian or classical conditioning.* A child associates the sights of a German shepherd with the memory of having been bitten. The result of the bite (or conditioned response) is a rush of adrenalin and a panicked reaction.

2. *Operant or Skinnerian conditioning.* An infant's crying brings Mama and a bottle. This action and its pleasant consequences are associated. The infant cries again when it wants something from Mama.

3. *Avoidance conditioning.* The child, bitten by the dog in one block, follows another route to school. This type of learning is particularly robust and very difficult to "unlearn."

The child does not learn that the dog is now friendly because she avoids another meeting.

4. *Perceptual learning*. This aspect of learning concerns the increased capability of the organism—boy, girl, or animal—to observe or *attend to* features of the environment. An infant cannot distinguish his mother from other people until he is several months old. Wine tasters can discriminate subtleties of wine and identify age and vineyard with extraordinary precision. Somehow a differentiation of features takes place with continuous experience. Since a particular wine may be identified through a variety of features that are "associated" together, some associative principle may also apply but, whatever the mechanism, differentiation must logically be prior to association. The ability to differentiate occurs partly as a result of neurological growth as well as from processes akin to adult learning. Through this, the ability to remember is also linked to neurological growth.

5. *Perceptual adaptation*. A variety of effects are included under this label. A common example is the adaptation to tinted sunglasses—the colored effect disappears after a short period of use. A number of experiments have been performed with distorting spectacles. Subjects adapt so that they see "normally" even if the spectacles actually provide them with an upside down picture of the world. This adaptation takes a couple of weeks. What role does memory play in this process?

6. *Perceptual motor learning*. We learn complicated motor patterns such as riding a bicycle or bowling. We talk of coordination and timing and of watching what one is doing. Coaches and trainers can point out "errors" or awkward movements. Both the visual system and the muscle system of the body are learning separately but in coordination. The greater part of skill and perceptual–motor coordination is actually achieved without conscious interaction. Memory is almost built into physical action patterns.

Skill learning is perhaps a basic property of nervous structures. It points to a differentiation principle: Irrelevant and excessive movements drop out and the optimum movement emerges as a stable form or structure. Reinforcement does not appear necessary for this type of learning—improvement is its own reward. Childhood seems to be an optimal period in which to begin this type of learning in many areas.

7. *Episodic memory*. The most common kind of remembering is that of recalling some event or happening. Young children do not recall events as accurately as older children. They tend to add a great deal of imaginary material and often recall events in the wrong order. Piaget feels that children, even in the middle years, still lack operational schemata for representing past events.

8. *Factual memory*. This general but convenient term refers to our knowledge of the world around us, past and present. It is gained from indirect sources: books, people, maps. This is the kind of memory that identifies historical events and people for us—a very different identification than recognizing the familiar face at the store. Children go to school to learn many facts that orient them to their identity, culture, and geography. Nothing may be *done* with this knowledge once they pass an examination, but we do not know what the ultimate effects of such knowledge may be on actions and social attitudes.

Recent interest in memory has led to an examination of an individual's structures of knowledge. The emphasis is on how facts are organized in the mind. In a number of experiments, adult subjects have been asked to memorize a group of words. Psychologists found that the subjects would cluster these words, reorganizing them in the process in order to attach meaning to the clusters. The correct order of the words could be learned but it took longer.

Children show clustering effects, but their ability to classify is weak and they do not cluster as much. Their growing ability in the area of memory is clearly related to a number of factors; mastery of the ability to classify may be one of them.

9. *Tacit knowledge*. The notion that we know more than we can say has recently become important to theorists seeking models of knowledge or language acquisition. Factual learning reflects more than simple associations between items. The items carry meanings and implications that must be actively incorporated into a consistent structure or, in Piagetian terms, schema. A fact such as "Napoleon marched into Russia," presupposes knowledge of a number of other facts: Napoleon is a person; Russia is a country. The fact increases its scope of meaning to those who know that Russia is a long distance from France, Napoleon's own country, that France was attacking Russia, and that winter conditions prevented Napoleon from achieving his goals in Russia. The importance of tacit knowledge is that it is taken for granted; it is learned implicitly and unintentionally, yet it forms the scaffolding for explicit learning.

Children generally cannot learn efficiently by rote. Like adults they seem to learn material best that occurs in a context—there must be some "readiness" to learn facts. However, rote rehearsal does produce some learning and can be of real use with languages or with other originally meaningless or disconnected material, which must be mastered before a basis of association can be formed.

Research evidence indicates that basic memory processes do not change greatly after the age of four. The strategies and structures upon which these processes work, however, do seem to change as the child acquires greater knowledge and greater ability to organize and to work with that knowledge.

Reading

Reading readiness implies that children are at a maturational point where they can learn to read. At the most basic physical level, readers must be able to coordinate their muscles in order to hold a book and move their eyes back and forth across a page. Perceptual skills must be fine enough to allow children to distinguish letters and to recognize words—very young children, for instance, tend to reverse letters. The backwards S is a "cute" characteristic of small children; consistent reversal of letters such as the d in Dad (to b), however, can cause great difficulties.

Reading brings together many of the cognitive skills learned in early childhood. Children must be able to grasp concepts, relate symbols to sounds, and put them in order before they can read. The graphic form of words in a culture indicates that this may be the source of some early difficulty in English. Children learning a pictorial language (each word is represented by one symbol rather than a combination of letters) such as Chinese have less early difficulty with reading than American children, who must learn a separate symbol for each letter of an alphabet then arrange the letters to make sounds and words.

Reading involves the ability to work with language. Children's vocabularies must be large enough to encompass the meaning of the material they are asked to read. Grammatical constraints must be recognized including ordering of words in a sentence and their effects on meaning. Orthographic rules, or how words are generally spelled, must be learned for children to recognize and reproduce words. Accepted stress and intonation patterns that convey meaning (the voice rising at the end of a sentence to indicate a question) must be known before the punctuation representing them can be learned. Children need to learn to distinguish homonyms, or words that sound the same but are spelled differently, and synonyms, different words that carry the same meaning, and then use them. Ultimately, familiarity with these linguistic structures in combination with other cognitive and developmental skills enables children to "decode" the letter strings and punctuation on a page into meaningful communication.

A number of experiments have been made that relate the importance of awareness of sounds and meaning of the language in learning how to read. One of the single most important difficulties in children who can't read is their trouble in relating the sounds of spoken language to printed symbols. For instance, many children with reading difficulties cannot rhyme. Psychologists have found that they cannot talk in pig Latin or take boy away from *cowboy* to make *cow*.

How a child learns to read remains mysterious. Most children eventually learn. Some do not. Certain factors beyond the developmental ones seem to aid reading. A home environment in which warmth and security are present and in which the written word is valued shows a correlation with reading success. However, this is not necessarily a causal factor in reading ability; this kind of background is also correlated with general mental development.

Because reading is such a basic and important skill, it has occupied the attention of educators for decades. They have not, however, reached a general agreement as to how and when reading should be taught. It has been taught to different age groups, and a number of strategies have been evolved to teach it. Most teachers would say that there are some children who do not seem to respond to any method and there are some children with whom almost any strategy would be successful.

Study aids

Review Questions

1. Johnny can't read. Based on the material in this unit, do you think he would be good at making up rhymes?

2. Draw four lines that go through every dot in the diagram without lifting your pencil from the paper.

```
    •       •       •


    •       •       •


    •       •       •
```

a. Why would we tend to call a solution to this problem "creative thinking?"

b. Would the creative person who solves this problem be more likely to use *convergent* or *divergent* thinking?

c. Can you assume that this creative person is also intelligent? Explain your answer.

3. "The light turned green, he pushed the pedal to the floor, and the Jaguar shot forward."

a. What kind of facts would you need to make sense of this sentence?

b. What do we call these background facts?

c. Is any *perceptual motor learning* implied in the man's actions? Explain.

4. What is *reading readiness* and how is it related to maturation?

5. Can you give an example of egocentric speech that also reflects egocentric thought?

Questions Pondered by Psychologists

Most people perceive this picture as a three-dimensional cube.

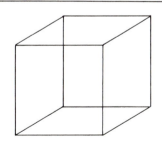

Why is it not perceived as a "flat," two-dimensional pattern?

Most of us reconstruct many events—such as those during childhood—from stories, pictures, and other outside sources of information as well as from memory. All of these are incorporated into what we perceive as remembering. Can they ever be sorted out? What are the implications for historians or for witnesses of an accident or crime?

Of Policy Matters and Public Interest

Methods of teaching art vary greatly. The Japanese and Swiss both train young children to draw carefully executed pictures with considerable emphasis on rules of perspective and a great deal of detail; critics say they are too constrained and "unchildlike." American children are allowed to be free in order to develop creativity; their pictures have few constraints and, according to critics, no form. The training that is needed to produce great artists, if any, is still open to question.

21

Aspects of Intelligence

Student objectives

1. Describe some of the purposes and appropriate and inappropriate uses of intelligence testing of children.

2. Describe what standardized IQ tests, such as the Stanford-Binet and Wechsler, are designed to do.

3. Describe the range of influences on standardized test scores, such as birth order, cultural bias, testing setting, and so on.

4. Describe several current definitions of intelligence.

Assignments for this unit

1. Read pages 307–332 and 440–448 in *A Child's World*.

2. View Program 21, "Aspects of Intelligence."

3. Read Overview 21 in this chapter and review the study aids. Note that theories of intelligence, well covered in both the text and program, are not reviewed in this summary.

4. Read Constructing Intelligence Tests in this chapter (optional).

In today's school, IQ is almost as commonplace a statistic as height or weight. Most children take an IQ test at least once. If general aptitude and achievement tests are included in this category, most take many such examinations over the course of a school career. If they score well, the school environment is favorable to them; if they do badly, the scores may provide information that determine the course of their lives.

This ubiquitous test does much more than provide an indication of certain abilities; it shapes a common perception of "intelligence." Most people think of others with high IQ's as having greater—not necessarily different—intelligence than the average. This, however, is a rather barren view of intelligence. As we have seen in earlier chapters, the concept of intelligence is far more complex than indicated by the linear comparison of an individual with the population at large. Furthermore, there are many factors such as wisdom, good judgment, motivation, creativity, or personality that an intelligence test might not measure, but that would make an individual superior in the actual performance of tasks throughout his or her life. The concept of intelligence as reflected by a test score is not nearly as simple and straightforward as many believe.

Students of behavior generally distinguish between two broad classes of actions: instinctive and intelligent. Animals that are lower on the evolutionary scale exhibit more automatic or "fixed-action patterns" than higher animals. A "fixed-action pattern," triggered by an event or series of events, is stereotyped in execution. Nest-building, for instance, a highly complex set of actions, is not considered a sign of intelligence in birds. Each species builds nests of only one design using only certain materials; if it is impossible to follow the pattern, the nest-building is upset. Intelligent behavior in such circumstances would be exhibited if the bird adapted to the new conditions. Instead of adapting its behavior to the conditions, however, the bird ceases construction or keeps on building a failure of a nest.

The ability to adapt is an important component of intelligent behavior, but this phrase has a very broad meaning. Some more specific indicators are often associated with intelligence:

1. *Modifiability of action*. A trial of other strategies or actions if a first attempt to reach a goal fails.

2. *Practical insight*. The ability to perceive different combinations of actions, materials, or ideas that might provide a solution to a problem or the achievement of a goal.

3. *Multiple solutions*. The ability to recognize that a problem may have more than one "correct" solution.

4. *Simplicity of solution*. Efficiency in adaptive behavior. Alexander the Great demonstrated this kind of adaptive behavior when he decided to cut the Gordian knot rather than try to untie it.

It is important to note that these types of behavior do not constitute a definition of intelligence. They merely indicate some of its aspects. Many definitions of intelligence exist—some of them are outlined in the textbook and television program—but none has yet been found to be universally acceptable for all purposes.

One of the earliest systematic approaches to the idea and measurement of intellectual ability originated in France. In 1904 the French Minister of Public Instruction established a commission to study the problem of identifying mentally retarded children. The purpose of this directive was to be able to give these children special instruction. It also provided the incentive for Alfred Binet, the director of a psychological laboratory at the Sorbonne in Paris, to publish his first intelligence test. The test was designed not simply to detect retardation, but to rank all children on a scale of relative intelligence.

From the beginning Binet and his later colleague, S. B. Simon, wished to measure "general intelligence" rather than any combination of specific abilities. They were concerned with the ability of children to function in all situations to which they might be exposed in a school environment.

Throughout this and subsequent revisions of the test, Binet maintained the primary purpose of scaling children on general ability. In order to do this, he tested children in a variety of areas: (1) perceptual—visual coordination, discrimination of lines, detection of missing weight, and so on; (2) linguistic—identifying objects by name, naming objects, defining objects; (3) memory—for sentences, pictures, drawing from memory, remembering spoken series of numbers; (4) social—carrying out simple orders and replying to problem questions.

In a scale devised in 1908 Binet introduced two new factors. These were *mental age* and *age-graded test items*. The average child could pass the items designated for his or her age level. For each successive year these items were more difficult. A bright child would pass items designed for older age groups, and a slow child would not be able to answer the questions for his or her age level.

The mental age of a child is the age level at which the child performs. The Intelligence Quotient is the ratio of the mental age (MA) to the chronological age (CA). A simple measure is the formula

$$IQ = \frac{MA \times 100}{CA}$$

although most IQ scores are now calculated by computer using sophisticated statistical methods.

Alfred Binet's intelligence tests were brought to the United States and refined by Lewis Terman. The 1916 Stanford revision (named for the university where the work was done) of the Binet scale by Terman was very successful and widely used. The test was further revised in 1937 and 1960, and today the Stanford-Binet is generally taken as a standard against which other IQ tests are validated.

It is important to approach the Stanford-Binet and other IQ tests with the understanding of the purposes for which they were devised. Alfred Binet originally wished to predict the probable scholastic performance of school children. Today's Stanford-Binet, stressing verbal ability, provides a good measure of a child's probable achievement in the school system, but it can mask certain problems that are not specifically related to potential, such as those caused by a negative home environment. Furthermore, a child may have abilities that are not indicated by this particular test.

Partially because of these deficiencies and partially because of the different approaches to the concept of intelligence, many other IQ tests have been devised and a number of them are widely used. David Weschler designed an intelligence test for adults in 1941 that has become one of the best known. This test provides separate measures of verbal ability (verbal IQ) and visual–spatial skills (performance IQ). Weschler had recognized that clinical patients often exhibited problems in performance tests but had virtually unimpaired verbal ability, and vice versa. These tests, now with editions for children as well, can detect abnormal discrepancies between these general skills and thereby aid in the diagnosis of brain injuries or similar problems. In general testing, the two sections provide an opportunity for those who are able in performance areas but not so skilled in verbal activity to demonstrate their abilities.

Factors other than the design of the test influence the scores that test-takers achieve. One of the most obvious is the testing situation. A room that is hot, loud noises, or negative attitudes from a test administrator can all affect a child's performance. Similarly, if a child slept poorly the night before a test, was upset by the test situation, or had caught her right hand in a car door the previous day, the scores would probably be affected negatively. Extensive experience in taking these tests or a very supportive administrator may, conversely, result in improved scores.

A more subtle influence is provided by cultural factors. "If you wished to reach a town a mile away, would it be quicker to take a tonga or a bullock cart?" Most American children and adults would have no idea; a child from India would know that tongas, pulled by ponies, are faster than bullock carts. It has proved almost impossible to devise an intelligence test that transcends cultural boundaries for these tests measure what has been learned and then uses that as a basis for what might be learned in the future.

A final, if currently unresolved, factor that may influence scores on intelligence tests is birth order. Some recent research indicates that first born and only children score higher on such tests and generally seem to perform better in both school and careers than their younger brothers and sisters. A possible explanation for this is that first-born and single children spend a greater proportion of their time interacting with adults. It may be that the intellectual quality of this interaction is superior to a young child's interaction with slightly older brothers and sisters. It is an interesting point, but one that is not yet resolved.

The first two influences point to some of the abuses for which IQ tests have been criticized in recent years. Generally this criticism falls into two categories. The first is that the tests are constructed or administered in such a way that they may represent unfairly the ability of the children taking them. The second is that they are being used in inappropriate situations for inappropriate purposes. It may be unwise to judge a child's intelligence solely on the results of one or even more of these tests. The test measures only what it is designed to measure and a corresponding definition of intelligence is "what you are testing for." Superior verbal ability may not mean superior ability to perform well in science, in a career or, in some cases, even to do exceptionally well in school. In spite of this, people often feel that high scores on a test such as the Stanford-Binet do indicate superior ability in all areas. This feeling, in turn, may affect a child's future performance. Experiments have indicated that some teachers, told that students had superior IQ scores, went out of their way to aid these students so that they actually did perform better than the majority of the class. In reality, the students had not scored as well as the teachers believed. IQ tests have also been used to argue that certain groups of people are mentally inferior, bringing what may very well have been an educational and cultural difference into the area of race and public policy.

On the other hand it does not seem wise to discard IQ tests in spite of the problems associated with them. They can identify children who are underachieving and, by doing so, indicate areas where possible problems might arise or when enrichment is needed. They are more objective than personal evaluations of a child's ability by teachers, who, like anyone else, can be influenced by their general feelings concerning a child. They can be used to diagnose some kinds of mental disorders or malfunctioning, as in the Weschler test and, as they are intended to do, they provide a very good indication of probable performance in certain areas of living and learning.

Study aids

Review Questions

1. In what area has the Stanford-Binet been particularly successful in predicting performance?

2. John, whose chronological age is 12, is found to have a mental age of 15. What is his IQ?

3. How do the different structures and content of the Stanford-Binet and Weschler affect test results?

4. Compare or contrast aptitude, achievement, and intelligence tests.

5. What is the meaning of correlation?

6. Can Guildford's theory of intelligence be called a refinement of Binet's? Why or why not?

7. An egg is to an omelet as cheese is to (a) fondue (b) cow (c) casein (d) salad.
   a. What kind of knowledge would a question such as this require?
   b. Could it be unfair to a student on the basis of sex, class, or culture?

c. Can you devise a test question that would be biased in favor of middle class males?

8. What, besides a child's intelligence, can influence test scores? What implications might your answer hold for the administration and evaluation of test scores?

9. Are there any problems in testing very young children? Do we have any statistical measures that might indicate the extent to which these problems could affect test results?

Questions Pondered by Psychologists and Psychometricians

How stable is intelligence over the life span? To what extent are differences registered on IQ tests administered at different ages a function of imperfections in the tests rather than changes in intellectual capacity? What relation does intelligence as measured by IQ tests have to the ability to achieve in later life?

Of Policy Issues and Public Interest

Are the injustices caused by the misuse and bias of some intelligence tests great enough to halt such testing in the schools? Some people argue that the tests are weighted against certain groups in our society and that people through the child's future will pass judgment on the basis of that test score. If a child has a bad day, is frightened of tests, or responds badly to a particular test, he or she may be marked for life. Others reply that IQ tests are successful in identifying certain types of problems and indicating what a child's school performance should be. Occasionally they add that performance in this society demands some of the same skills as performance in school. Many point to the need for more careful administration and use of IQ tests. Several states have banned the use of IQ tests if a school does not have written permission from children's parents.

Construction of the Stanford-Binet

Test limitations and possible interpretations of test scores depend on a careful study of the method of construction of the intelligence test being used. The actual formulation of such a test is the work of psychometricians and is extremely complex, but we can examine some of the principles in relation to a particular IQ test, in this case the Stanford-Binet.

Lewis Terman believed that intelligence was the capacity to think abstractly. This general definition allowed him great scope in the selection of test items, and he included several areas of mental operations.

Test items were made up by experienced testers and tried out many times. Statistical techniques were used to indicate the relationship or *correlation* between success on an item and the overall score on the test. If an item were failed too often by children who had otherwise high scores, it was discarded. This procedure ensured that only those items that contributed consistently to the test score were included. This process of correlation maximized the *internal validity* of test items.

Terman then refined Binet's age levels. At each age level the test items were progressively more difficult. Within each age level, items were selected that could be passed by an arbitrary 60% of the population and that were intended to indicate present ability independent of schooling or special experience.

The test items were also *standardized*. They were tested on a large sample selected at random, which included children of school age from all social classes and regions. If girls and boys consistently scored differently on any test items, the items were dropped. This process, however, was not carried out for racial or cultural minority groups. The original standardization sample was primarily white so an implicit cultural or racial bias could have been built into the Stanford-Binet test. Ideally the composition of a standardization sample should match that of the population on which the test is to be used. As we have mentioned, the sample originally used does not always now reflect the population taking the test.

The purpose of standardization is to establish scoring norms or the proportion of children at different ages passing or failing any item. Once these statistics are down, the performance of any single child can be compared to the average for the test group and the IQ score computed.

Most people score near the average mark with relatively few scoring very high or very low. The result is a bell-shaped curve that tells us IQ is distributed randomly in the population. The "bump" at the low end is due to a specific genetic factor that causes Mongolism.

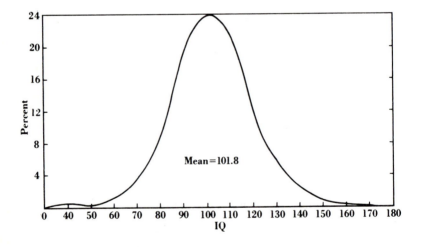

The main purpose of establishing and administering an IQ test is to provide an independent and reliable estimate of an individual's ability to do well in the mental tasks stressed by the test. In practice the validity of a test is judged by its ability to "predict" achievement in some sphere of life, chiefly scholastic. If a test has good *predictive validity* for school performance, a group of children ranked according to their IQ should have and maintain about the same rank in their school work. The test could then be described as "correlating highly" with school achievement. The Stanford-Binet was designed to do exactly this.

22

The Child's Personality

Student objectives

1. Discuss the complexity and interrelationship of factors affecting personality development, including psychoanalytic and psychosocial theories of development.

2. Discuss the development of some personality components important to this period: self-esteem, independence, and achievement/competence.

3. Discuss the importance of separation from parental figures and the increasing significance of peers and the childhood society in this period.

4. Discuss the development of self-concept and describe how such factors as prejudice and sex typing affect its development.

Assignments for this unit

1. Read pages 481–512 in *A Child's World*.

2. View Program 22, "The Child's Personality."

3. Read Overview 22 in this book and review the study aids. The overview in this unit primarily expands objective 2.

4. Read Defense Mechanisms in this chapter (optional).

The transition from Piaget's preoperational period to the stage of concrete operations in mental development is accompanied by a similar transition in personality development. Freud calls the middle years of childhood a *latent* period. According to the psychoanalytic theory, there is a reduction of overt sexuality after the age of five or six, allowing the child to grow and develop in other ways during the next years. In Erik Erikson's psychosocial theory of development, middle childhood is also a period of consolidating attitudes and laying the foundations for future participation in adult society. Erikson holds that children of this age bring work into their lives, looking to their activities for a sense of achievement and competence and to enhance their self-esteem.

Self-Esteem

The middle years are a time when the forces of socialization become particularly important to children. Their abilities, potential abilities, and interests are influenced by the social forces around them. Sex-role differentiation becomes clearer as boys choose to play with other boys and girls with girls. In interacting with their peers, children increasingly learn to take other people's points of view. This process heightens their awareness of others' opinions concerning their status and popularity. The result is often a circular feedback process for they respond to these opinions. Their response, in turn, shapes others' perceptions of them. Children with high self-esteem or a positive perception of themselves and their attributes thus gain a double advantage: they socialize easily and make friends; their popularity, in turn, helps them maintain their positive image of themselves. It is not easy to disentangle the cause of positive or negative self-esteem from its effect.

The experience of success is another important factor in self-esteem. The drive toward autonomy and competence that characterizes this period tends to focus children's attention not only on their efforts but on the results they achieve. Failure is discouraging and frustrating. Failure can be a positive experience in that it may force a child to appraise his abilities more realistically, but too much failure too soon, particularly if it is associated with an already low degree of self-esteem, can establish a vicious cycle. Those children whose self-esteem may be low because of negative stereotyping (Unit 16) are especially vulnerable to this kind of cycle.

Adults can, of course, exert a strong influence over a child's sense of worth. Negative interaction in the form of excessive punishment, disapproval, or neglect of a child's productive efforts can be devastating to future creativity. Parents, more than anyone else, provide the growing child with an orientation toward success or failure. If the child's attitudes are positive, those attitudes can provide a buffer against failure. If they are negative, only constant positive feedback from other sources can compensate for the child's lack of a sense of self-worth. According to Alfred Adler, an early disciple of Freud, many successful people pursue success compulsively in order to overcome feelings of inferiority that were acquired early in life.

The Move Toward Independence

The middle years mark an initial break with the family. Children in this period tend to form strong relationships with people outside the immediate family group. Autonomy, however, is a relative term. As children gain a measure of independence from their parents, they become more dependent upon the opinions of peers. The time they spend with children their own age increases rapidly as they move from the preschool period into middle childhood, and the time they spend with parents decreases. Leon Rappaport, a psychologist who has done much work in this area, suggests that peer relationships are advantageous to the growing child in four respects:

1. *Social support.* The peer group allows the child to share values and build dreams. Parents, on the other hand, can come to be regarded as restrictive and tyrannical, in fantasy if not in fact. The symbolism in Jack the Giant-Killer should be clear in this context.

2. *Models for imitation.* Friendships outside the family with children of roughly the same age provide the child with easy models for imitation. A slightly older child with special competence in one area or another can quickly find a following among younger children. Older siblings, quite naturally, often provide models for younger brothers and sisters.

3. *Social roles.* The active participation of the child in a variety of social situations enables her to adopt different social roles. She can also learn to relate flexibly to other people's roles. In one situation she may be a follower; with another child she can lead. Children's gangs generally have a role structure similar to those of authoritarian adult groups. The strong role differentiation of childhood society can provide a child the security derived from a strong sense of belonging.

4. *Standards for self-evaluation.* Security also comes from knowing where one fits in. Not everyone can be chief, but everyone can have a useful role. Self-evaluation here implies a desire for assessment by the peer group, assuming, of course, that acceptance has already been gained. (Rejection or expulsion from a group is a traumatic experience for anybody.) Peer interaction helps the self-concept develop along realistic, stable lines, enabling children to avoid severe mismatches between their self-concepts and their actual roles, interests, and abilities. The importance of peer interaction, however, does not supersede the importance of the family. The move away from the family that we see in middle childhood is only a tentative step; a child's parents will be of great influence in all areas of development for a number of years to come.

Review Questions

1. What role does sexual latency play in this period according to Erikson's psychosocial theory of development?

2. What special significance might the saying, "A man is known by the company he keeps," have during the middle years of childhood?

3. Define the following:
   a. Repression
   b. Reaction formation
   c. Projection
   With what body of theory are they associated? What are they?

4. Describe an experiment dealing with the influence of adults on peer group dynamics.

5. What is self-concept and why is it important?

Questions Pondered by Psychologists

Is the need for achievement/competence unique to humans? Do animals have it? One psychologist (David McClelland) argues that the desire for achievement—inculcated in childhood—is linked to economic development.

Defense Mechanisms

Healthy "normal" children in the latent period of the middle years will show few inner conflicts, tend to relate easily to teachers, peers, and other adults, and enjoy loving relationships with their parents. The ideal child, however, is rare. More often children display a variety of anxieties, fears, learning difficulties, and problems in adjustment. An unsettled home environment, for instance, may hold back emotional development and make school a stressful experience. In such circumstances undesirable adaptations arise to protect the child against his anxiety. In Freudian theory these are called the *mechanisms of defense*.

Defense mechanisms are intended to prevent the ego from being overwhelmed by the unconscious forces of the id and to reduce the guilt brought on by the superego. Defense mechanisms represent stop-gap measures that keep unconscious impulses out of awareness. Unfortunately this denies the ego the opportunity to deal with them rationally. If the defenses are too successful—and they often endure a lifetime—they can have a detrimental effect on personal growth.

The familiar term *repression* refers to the relegation of an unpleasant thought, memory, or drive to the unconscious. There it may stay, unless unearthed by psychotherapy, but its effects do come to the surface. These may influence emotional life and behavior, especially if the repressed material concerns sexuality. *Suppression* is a milder form of repression and refers to a temporary removal of thoughts to the unconscious. Suppressed material is less loaded emotionally and is capable of coming into the consciousness eventually to be assimilated into the ego.

Projection is another well-known term. Freudians use it technically to mean the attributing of unconscious (perhaps repressed) intentions or drives to another person. If a daughter cannot accept her hostility toward her mother, she may say (and think) her mother hates her instead.

Reaction formation represents an emotional response to unacceptable feelings. The daughter may persuade herself that she has no hostile feelings toward her mother at all.

Displacement is a type of symbolic release of psychic energy. If an emotion cannot be directed toward the true object, it is "displaced" onto another object. Instead of hitting her mother, the daughter may beat up her baby brother.

Rationalization is perhaps the commonest and most readily observable defense mechanism. When the superego indicates that an action is likely to result in guilt feelings, the person presents himself or others with apparently rational reasons for carrying out that action. The rational argument is often sustained by *denial* of objective facts or feelings. Denial is a simple mechanism whereby the person avoids facing the issue directly, yet does not deliberately intend to deceive. "No, I *love* my baby brother. I hit him because he was going to make a mess with the crayons and I didn't hit him *very* hard."

Finally, *sublimation* refers to a transmutation of psychic energy into more refined and creative forms. It may be the most useful mechanism of defense in terms of productivity. Freud thought that intellectual and artistic production was fueled by the frustration of erotic desires. Thus, the latent period of the middle years is supposedly the time for directing the child's energies toward developing skills and interests. Later when biological change in adolescence generates new forces, the sexual energies can be more easily sublimated because of previously formed interests and habits.

23

Moral Development

Student objectives

1. Discuss the concepts of moral development and conscience with reference to Kohlberg, Piaget, psychoanalytic theory, and social learning.

2. Describe the characteristics of moral thought in infancy, preschool, middle years, and adolescence.

3. Discuss the factors that affect moral development in children.

Assignments for this unit

1. Read pages 422–437 in *A Child's World*.

2. View Program 23, "Moral Development."

3. Read Overview 23 in this book and review the study aids.

Moral development is a subject that is usually associated with philosophers and theologians rather than psychologists. It is a difficult area, for before moral development can be studied, there must be some agreement about the definition of *moral behavior*. This is an issue of unending complexity and, indeed, may be one of the reasons that psychologists have given moral development little attention until recently.

In order to study moral development, psychologists have begun to develop and work with a concept of morality that deals with social and cultural standards. In this approach *moral* does not refer to the content of actions but to the individual's and society's perceptions of them. A cannibal would not be judged immoral because he ate human flesh if cannibalism were an accepted practice in his tribe. Whatever a psychologist's personal feelings might be, the cannibal's moral development in psychological terms depends on what he, Sir Cannibal, thinks about eating people, not upon what the psychologist thinks. Since much of the work on moral development has been done in our culture, studies refer to standards most of us hold. You should be aware, however, that few psychologists would claim that the things we feel are of the highest moral order are universally accepted as such.

Infants do not behave in a manner that can be classified as *moral* or *immoral*. Their behavior is *amoral*, or neutral with respect to moral considerations. As infants become children, behavior becomes related to morality. Three-year-old Tommy hits his baby sister because he wants her toy. Four-year-old June hides the cookie she stole behind a vase. Six-year-old Jenny lies about riding her bicycle in the street. Each of these actions illustrates a different violation of our rules of morality. Tommy expressed his agression with violence toward a helpless infant. June was deceitful. Jenny disobeyed her parents and lied.

These moral rules refer to social conduct. They regulate how people behave toward one another. Violation of these rules often leads to negative consequences because other people retaliate or punish the offender. Retaliation or punishment is normally justified by an appeal to a higher authority such as the law or religious standards of conduct.

For most of us, an important component of moral behavior is intention with awareness of the rules. Infants, therefore, do not show moral or immoral behavior because they have not yet been taught the rules of conduct. If Natalie stumbles and drops a glass, adults are less likely to consider her behavior immoral than if she threw the glass on the floor in a tantrum. If Tony has not been told the cookies are for the church bazaar, his eating them is not judged as harshly as it would have been had he known he shouldn't touch them.

Recognition of intention and knowledge of rules, however, are both adult judgements for the perception of moral standards comes gradually throughout childhood. At an early age, moral notions are absent altogether. If something is "wrong," it is because somebody said so. If Heather breaks some cups accidentally, it is still wrong, and how "wrong" it is depends on how many cups are broken and under what circumstances.

Later, about the time children begin school, morality consists of following rules. The rules can come from parents, teachers, the President of the United States, or God and are not questioned. Immorality occurs when they are broken.

Finally, in adolescence, most individuals begin to think about *moral principles*. These are the social values embodied in the rules, laws, and social customs and might include such things as individual rights and reciprocity in human relationships.

Why and how does this development occur? This unit examines four theories concerning moral development: Piaget's, Kohlberg's, psychoanalytic, and the social learning or behaviorist.

Piaget believes that progress through the stages of moral development is the result of cognitive growth—a combined effect of biological maturation and experience. The preschool child is not yet mature enough to understand other points of view. Her *egocentrism* limits the extent to which she can comprehend moral laws or rules. In the stage of concrete operations, she is able to see that rules are a necessary means of governing conduct between people. Finally, in the stage of formal operations, moral principles can be grasped as higher order values. Through all of this the importance of intention is recognized as children become better at perceiving and understanding others' motivation.

Kohlberg elaborates Piaget's stages (see text, pp. 431–432). He finds two substages within each main stage, but because he formulated these substages so recently, we still have little experimental evidence concerning the validity of his theory. General support for the Piaget–Kohlberg emphasis on the role of cognitive development in moral development, however, is provided by the correlation of IQ with these stages. Brighter children tend to be more advanced in the types of moral judgements given in the text.

Social learning theory, derived from the *behaviorist* approach, is based on the concept of modeling, or *imitation*, as the chief means through which a child develops moral standards. Children imitate specific bits of social behavior. If these are acceptable, they are rewarded with approval or with the responses they desire. If not, disapproval or punishment inhibits further expression. Language is important in this theory in that it helps to categorize social situations and people and provides a means of controlling attitudes and actions (see Unit 22).

The psychoanalytic approach to moral development is concerned with its emotional and motivational aspects. Moral behavior is a means of avoiding guilt. Its foundation lies in *identification* or the child's tendency to assimilate the positive attitudes and traits of a person he loves. Identification is not literally imitation of the loved person's behavior (as in social learning theory). It occurs initially during the Oedipal phase when the child does not treat the same-sex parent as a rival but accepts the parent's authority. The instructions and commands of both parents become incorporated into the *superego* so that the child can regulate his own behavior even if the authority figures are absent.

At the point when violations of the superego cause guilt feelings and anxiety, true moral behavior appears. The child can now exercise a choice between giving in to the instinctive desires of the id and doing what he wants or doing what he knows is right.

A well-integrated person does not have conflicts between the id and superego, but people who are this well integrated are rare. On the other hand, overwhelming dominance of one or the other can cause difficulties. Severe parental training may produce a strict superego in a child. Overly high standards of moral behavior will restrict behavior and personality and cause guilt over the tiniest infractions of the inner law. A weak superego will create different problems. The child will exceed social bounds and collide with externally imposed controls more often.

In all theories moral development is influenced by the environmental standards surrounding the child. The kinds of behavior required and the methods through which they are enforced are important. Parents, for instance, who punish a child without providing a reason beyond "that's bad," are not encouraging her to think about her actions. Discipline based on power seems to be correlated with a slower advance in moral thinking than *rational discipline* or discipline associated with explanations concerning a transgression.

Peer interaction is important to moral development. By relating to others, making friends, and engaging in cooperative activity, children learn to enter different roles and empathize with each other's intentions and desires. Children with more social experience tend to display a higher level of moral reasoning.

Parental affection is also necessary for moral growth. Many problem children are the result of lack of parental concern rather than parental control; an affectionless environment has devastating consequences. Affection need not come only from parents. A number of experiments indicate that children will disobey instructions less often in the presence of an affectionate, interested person than in the presence of a figure who does not relate to them.

Finally, we will note a complicating factor in the study of moral development that applies to adults as well as to children. There are very often three sets of moral standards; sometimes individuals are aware of them and sometimes they are not. There are the moral standards individuals say they believe in; those they believe in; and the ways they act. Very often these are distinct; sometimes they are not. The separation of moral attitudes and actions poses problems in the assessment of moral development.

Study aids

Review Questions

1. George Washington, hoping to surprise his mother with a load of firewood, chopped down his father's favorite cherry tree by mistake. Based on the material in this unit, what kind of punishment might be recommended for this act by:
   a. A child of four
   b. An adolescent of fourteen
   What explanation might be given by each for his or her answer?

2. In Piaget's stages of moral development where would each of these individuals be?

3. How would the explanation of a social learning theorist concerning the difference in the four- and fourteen-year-olds' judgements of "fair punishment" differ from Piaget's?

4. Describe the evolving relationship of the ego, super-ego, and id in the psychoanalytic approach to moral development. In these terms, what might be a possible description of George Washington's action and his subsequent remorse?

5.   What evidence is there to indicate that Piaget's stages of moral development are not universal?

6.   How does peer interaction affect moral development?

7.   What is the relationship between Piaget's and Kohlberg's theories of moral development?

Questions Pondered by Psychologists and Philosophers

Does the basis for moral development go beyond the need for rules of social conduct? What should the relationship of psychology to religion be in this matter? At what point should a child or adult begin to be held responsible for his or her actions?

Of Policy Matters and Public Interest

During the peak of the demonstrations against the Vietnam War in Berkeley a number of Berkeley area residents were polled concerning their beliefs in the right to freedom of speech and in the right of nonviolent demonstrators to protest the war. The interviewees were overwhelmingly in favor of freedom of expression and opposed to allowing the antiwar protests to continue. On the whole, they did not perceive a discrepancy between these two positions.

24

Aspects of Socialization

Student objectives

1.   Recognize the variety of influences on child growth and development that stem from social organization and culture.

2.   Compare and contrast different social classes' expectations for children's behavior.

3.   Identify at least five agents of socialization of children.

4.   Identify ways in which attitudes and cultural values are transmitted to children.

Assignments for this unit

1.   There is no text assignment for this unit. Material relating to the influence of class and culture has appeared throughout this course, and you may wish to review some of it. It includes pages 203–204 (mental development), pages 302–307 (language development), pages 350–353 (sex-role development), pages 467–471 (socioeconomic status and education), pages 518–520 (relationships with parents), pages 526–533 (families

and child rearing) and 593–595 (vocational aspirations).

2.   View Program 24, "Aspects of Socialization."

3.   Read Overview 24 in this book and review the study aids.

4.   Read Some Aspects of the Process of Socialization in this chapter (optional).

Overview 24

A young American G.I. was guarding a field used for target practice in Japan in the 1950s. Suddenly he spied what appeared to be an old woman, gathering the used shells. This was forbidden for reasons of safety and security, so he waved her away. She looked at him and came toward him. He shouted and motioned again. Still the figure came. Nervous, for this was his first assignment abroad, the G.I. pointed his gun at her and pulled the trigger. The old woman died before they could get her to the hospital.

At the hearing that followed, a cultural difference emerged. The young American had waved his hand at the woman in a shooing motion while he shouted. She, of course, understood no English and that motion, in Japanese culture, meant "come here."

Not all cultural differences result in such tragedy, but neither are all such differences so obvious, particularly when they occur in the context of one society. This unit will look at social and cultural differences within American society and at the ways in which a growing child is influenced by them.

In a sense all Americans belong to one society. We share a common language, participate in a common economic and political system, and have a common cultural symbolism. When two Americans meet in a foreign land and one sighs "How I wish I had a hamburger and fries," most other Americans would understand, but few foreigners would unless they had spent time in the United States.

However, this does not mean that the United States is a homogeneous nation. Ethnic background, regional customs, and social class all interact to form a large number of subcultures. Some of these have a profound influence on the way children are reared and on the values and attitudes they form, for no child is reared in a vacuum. The bonds each child forms with others and the influence they, in turn, have upon that child have a profound effect on his life.

Socialization Agents

There are many influences on the process of socialization. In general terms, they include all those forces that affect a child's interaction with people around him. In this chapter we identify and discuss five types of socialization agents. These are: the family, schools, churches and other social organizations, the peer group (or other people outside the family) and the media (mostly television and children's books).

Family. Socialization starts with the infant's early interaction with the mother or primary caretakers. The mysterious process of *identity formation* begins early. According to Eric Erikson, it is the establishment of "basic trust" in people and the environment that allows the infant to venture forth and explore his world.

Researchers have found a relationship between social class and parenting styles. Middle-class parents frequently differ from working-class parents in the manner in which they discipline young children. Middle-class parents tend to *explain* the reason why the child should or should not do something. Working-class parents tend to *prevent* the child from doing undesirable things through commands or physical restraint. It is possible that this observed difference reflects a more general difference in child-rearing practices that may partially account for the better performance of middle-class children in such areas as IQ tests and preschool.

As the child grows and comes to interact with parents and schoolmates in more complex ways, the factor of "family" begins to imprint distinct characteristics on the child's behavior style. Attitudes of the parents toward school— positive or negative—are picked up and incorporated by the child. From an early age children show enthusiasm or negativism toward school work, which often reflects the home environment. The family—primarily the mother—also may provide the major and most lasting source of influence on the child's attitudes toward self and others. Racist stereotypes, for instance, have been observed in preschool children, indicating the impressionability of young children to social attitudes.

Bernard Rosen has related differences in achievement orientation to cultural differences in family background. He compared six ethnic and racial groups in four Northeastern states—French Canadians, Greeks, Southern Italians, East European Jews, blacks, and native-born white Protestants. The study showed that white Protestants, Jews, and Greeks imposed high standards of excellence on their children compared to the other groups. Cultural attitudes on the achievement dimension tend to be transmitted from one generation to the next and, he feels, at least partly accounts for the relative achievements of children from these backgrounds.

School. Once the young child comes into contact with other children of the same age, she finds there are considerable pressures toward socialization. These are increased by a school situation. By the age of six or seven the child learns that other children have their rights, that not everyone can talk at once, nor can every whim be satisfied. A fortunate child comes to enjoy school, likes the teachers, and strives for their approval. The adults and older children come to serve as secondary models for the child, and positive relationships at school can provide powerful incentives to develop self-restraint, responsibility, and diligence.

Such fortunate children are all too often from middle-class backgrounds. Unit 18 dealt with the problems of children from lower socioeconomic backgrounds in schools. Those whose families also belong to a minority ethnic group may have these difficulties compounded; in addition to being unprepared for school in either attitudes or skills, children may find they have no understanding of the values or culture the teacher expects them to share.

Churches and Other Social Organizations. Formal social organizations have varying degrees of influence on socialization. Unlike the institutions described in our other categories, participation in these groups is strictly voluntary and not universal. Nevertheless, they have a pervasive influence in setting moral standards.

Peer Group. The importance of the peer group has been discussed in several units. These groups may act to erase differences arising from class and culture (as in many second-generation immigrants who rejected their parents' cultures) or to reinforce them.

In examining the question of what makes a child popular in a peer group, researchers have discovered that certain factors seem to be particularly relevant. Popular or well-liked adolescents tend to be self-confident, well spoken and fluent, intelligent, athletic, quiet, and conscientious. Individuals who tended to be noisy, overly talkative, conceited or effeminate, were not popular. In general, both sexes seem to agree on the popular choices of each sex.

Popularity in the peer group of school children is related to parental attitudes toward the child: children who were well liked by their peers tend to have a tension-free home life with loving, nonauthoritarian parents. These children are also more satisfied with their home life than the average child. Prolonged absence of the father from the home seems to affect boys more than girls in their peer relationships, these boys display more aggression and have fewer friends. Good relations with the same-sex parent are important for peer-group adjustment, but this requirement seems to be more necessary for boys than for girls.

For reasons of obvious practicality, more research has been conducted on the effects of peer models on behavior than on parent models. Experimenters can present school children with all kinds of models, but it is not thought possible to use the child's parents in a controlled experiment. Researchers are generally confined to home visits to assess the quality of the parent–child interaction and the parents' personalities.

Television. Because of the apparent ease with which children can assimilate all aspects of behavior—desirable and undesirable—they see, the quality of television programming has come under scrutiny by parents, teachers, and public leaders. Although no clear evidence has linked television violence to crime, the relationship seems a likely one in view of the numerous experiments conducted in the laboratory.

Many studies have demonstrated that exposure to aggressive behavior in a child model causes that behavior to be reproduced in a similar situation. For example, a model was depicted as breaking and throwing toys. Later the subjects' play behavior with similar toys contained a higher incidence of such aggressive behavior.

On the other hand, the tendency to imitate a television gangster could be inhibited by those factors—arrest, parental disapproval—that inhibit crime or delinquency anyway. Thus, it could be that a child or adolescent who is in a situation in which he is likely to do something illegal may not be appreciably influenced by the television exposure, since other pressures in that direction are already great. Even if a direct link cannot be established experimentally between television crime and actual crime, however, the long-term effects on social values of exposure to violence must be seriously considered.

In another approach, some say that television mitigates the tendency to violence by allowing vicarious satisfaction in watching it. This theory stems from the ancient Greeks. The act of empathizing with a character in a play was supposed to make the viewer "live" the emotions of the character. Hence, instead of the audience acting out violent emotions themselves these are "expressed" vicariously through the actor. There's little evidence for this theory although it has been in existence for over two thousand years.

Positive benefits can be derived from television. Recent reports indicate that many children's programs are educating their audience to a significant extent. "Sesame Street" was designed to teach as well as to entertain. It seems to succeed in both these aims. Children who watch the program regularly do show gains in reading skills. Television does tend to foster passivity in the viewer (or learner), but this can be mitigated by innovative programming or peripheral activities. More serious may be teachers' complaints that young children expect school to be as entertaining as educational television programs.

Psychologists are now particularly interested not only in television but in all the influences that determine the social thoughts and attitudes of children. The answers are sparse, and we have discussed many of the possible influences in a descriptive manner. Generally, the more opportunity for social interaction with parents, siblings, and others in the home, the more a child advances socially. However, although a certain amount of social interaction seems necessary, the *quality* of this early social and linguistic environment may be even more important in developing cognitive skills.

Study aids

Review Questions

1. What agents for the socialization of children were identified in this unit?

2. How does the linguistic interaction between parents and children differ from class to class?

3. What is *socialization?*

4. What three stages of egocentrism in the socialization process are identified with Jean Piaget?

5. What role does the family play in the socialization of the child?

Question Pondered by Psychologists

What causes children to imitate some behaviors and not others?

Some Aspects of the Process of Socialization

At first, an infant has little idea of how to relate to people. He may smile at his caretaker but may also smile at a picture, his rattle, or his doll. Fond parents would like to think that his pleasure is a result of his love for them, but even they suspect it may have more to do with general well being.

In the fourth or fifth month, the primary caretaker—generally the mother—becomes the main focus of concern. Mother is now recognized and does seem to elicit genuine pleasure regardless of whether she brings baby food, diapers, or just herself. The baby can be distressed if his mother is absent for a long time, suggesting the formation of an attachment bond. Even children with multiple caretakers tend to have a primary attachment to one person. Severe psychological disturbance and even death can result from the separation of a child from the object of this primary attachment (Unit 9).

Once the importance of the mother as a person rather than as a provider of physical comfort is established, the child can learn to relate to other people around him. This is the beginning of the *socialization* process. To become socialized means to recognize others as human beings like oneself and to accept the values of one's society. As they grow older, children must learn how to behave toward peers and adults, how to gain approval and avoid antagonizing people, yet how to manage to express their own needs and seek personal satisfaction in relationships. To manage this complex of tasks requires (1) insight into others' intentions and desires and (2) insight into one's own intentions, desires, and behavior, recognition of whether they are appropriate, and knowledge about what responses can be expected. In short, the child develops *empathy*.

Empathy is thought of as the mysterious capacity humans possess of being able to "put oneself in another's place." Most animals probably do not possess this capacity, though cases of altruistic or *helping* behavior have been noted in animals such as whales, elephants, and wolves. The development of empathy may be a long-term process. Piaget claims that the child's egocentrism or incapacity to appreciate another's point of view is specially marked in three periods of social development. The first period of egocentrism includes early infancy up to age two; the second occurs about ages four to six, and the third period is in early adolescence.

Early infant egocentrism is a result of the infant's inability to differentiate her body from the rest of the environment. The infant has to learn, for example, that her hand belongs to herself, whereas the sides of the crib do not. By the active years of two to three, this separation has become clear.

In terms of social interactions, infants and children must come to pay attention to the emotional states and intentions of others. This begins at an early age. One psychologist, Paul Ekman, has studied the nature of emotional expression for many years. Ekman concluded that certain facial expressions exist in every culture and are probably biological in origin: anger, rage, amusement, and tenderness all share certain characteristic alterations of facial muscle tone. It is also known that children from the age of about four months pay more attention to a face than to an abstract pattern. At about this time they can maintain fairly consistent "eye contact" with the caretaker.

Facial expressions of the infant's early emotions are readily recognizable by adults and are presumed to be innate. It is not a big step to expect that infants are sensitive to *adult* facial expressions. Infants only a few days old will imitate other people's actions, such as putting out the tongue. Imitation of this type, useful in allowing infants to be able to express their feelings clearly, is an important part of the socialization process.

As children reach the preschool years, however, they begin to come into greater contact with other children of the same age and to learn the values of peer-group interaction.

Here a new type of egocentrism is shown—the inability of the child, who is now verbal and mobile, to appreciate the point of view of others.

In his early work on language acquisition, Piaget noted that young children will engage in what he called a "collective monlogue." A group of children, apparently in conversation, are really talking to themselves. They have not yet learned to *address* their conversation so the other will hear or understand it, nor do they *listen* when their playmate speaks. This absorption in their own ideas and newly formed linguistic skills is gradually replaced by more adaptive communication—the true dialogue.

This period of transition from preschool egocentrism to the stage of socialization reflected in the middle years represents a considerable challenge to psychologists. We have already reviewed the various theories of personality development in this period (see Chapter 22). Here the phenomenon to understand is how the child develops an awareness of social situations, the intentions, and the desires of others.

George Herbert Mead, an influential social psychologist in the earlier part of this century, proposed that the process of socialization involves the development of concepts of *self* and of *other*. In order to hold a conversation, for example, the child must be aware of whether the *other* is listening. Furthermore, the child must possess a sense of his *own* perspective as distinct from that of others. As the child grows up, the "other" is differentiated further than simply "mother" or "brother." "Other" also includes "family," "school," "scout troop," "team," "country," and so on. Each social unit can elicit definite roles. The well-known phenomenon of the Englishman dressing for dinner in the jungle can be related to the sense of what Mead called the "generalized other." The expectations of others are so strong in the Englishman's culture that he responds to them even when those others are absent or the response is inappropriate.

Egocentrism, submerged as the child develops a sense of "others," reappears as a result of the sudden challenge of adolescence. The emotional upheavals and the requirement that the young man or woman leave the home and establish independence imposes a new need for broader social perceptions. With the advent of formal thought (see Chapter 27) and the opportunity for wider social interactions, however, adolescents emerge from their initial absorption in their own problems, come to relate to the adult world, and finally mature into adulthood. All of these complex series of maturational and learning processes by which children come to function adaptively in a society are a part of the *socialization* process.

Social learning theory sees the end product of socialization in much the same way as Piaget but postulates a different process for its attainment. Socialization occurs through *imitation*. Children learn to behave appropriately by watching others. This imitation is a natural process and, indeed, can be observed in infants a few days old who will copy some facial expressions. But social learning theorists go beyond recognition of this as a phenomenon to treat it as a theoretical construct.

25

Childhood to Adolescence

Student objectives

1. Describe the characteristics of physical and motor development in the six- to twelve-year-old child and the adolescent, including factors that influence this development.

2. Identify the developmental changes that occur in pubescence (before the onset of puberty).

3. List the primary and secondary sex characteristics that develop during puberty.

4. Describe the emotional and social correlates of physiological and biological change that take place during puberty.

Assignments for this unit

1. Read Chapter 10 (pages 387–409) and pages 537–561 in *A Child's World*.

2. View Program 25, "Childhood to Adolescence."

3. Read Overview 25 in this book. Because the reading assignment is relatively long, this overview is short.

Overview 25

This unit is a transitional one, summing up the physical changes in the middle years and projecting physical changes into adolescence and adulthood. The range is great—imagine a child of six next to a young adult of eighteen—although in reality the changes from birth to six are just as great.

By the age of six, children have developed both fine and gross motor skills. They can hold a pencil and print, run, jump rope, and throw a ball. All of these activities will improve in quality as they grow into adolescence. By the time a young man or woman is 18 most of the maturational processes affecting physical skills have taken place, although specific actions may be improved through practice into late adulthood.

Physical growth is also complete for most people by the age of 18, although it follows no one regular pattern for everyone. However, most children grow throughout the middle years with little change in body proportion; then at the brink of adolescence they enter a period of profound change. At this time the growth rate increases, and youngsters mature sexually. Body proportions alter, sexual organs and breasts mature, and secondary sex characteristics appear. After about a year the growth spurt falls off as abruptly as it appeared. Both boys and girls continue to grow for some years thereafter but at a much slower rate. The rate continues to decrease until maximum height is attained. This occurs at about age 16 for girls and 18 for boys.

This growth is a *regulated* process much like the *homeostatic* regulation of internal chemical states according to Tanner, a world authority on physical development. If the growth of a child is retarded because of a temporary period of poor nutrition at some point, growth after that time is accelerated. This "catch-up" effect indicates that a mechanism of regulation is operating, almost as if the system had set an end goal of certain body dimensions.

Most aspects of physical growth parallel the growth curve for height—steady until the adolescent spurt, then decreasing thereafter—but some are exceptions. The reproductive organs—both genitalia and internal organs—show little evidence of growth until adolescence despite the overall increase in body size. At puberty these organs exhibit a relative growth rate greater than that of any other body structures. In contrast, the brain and head reach 75 percent of adult size by the time a child is four and grow very slowly thereafter.

Physical growth is thought to be determined primarily by genetically based mechanisms of maturation. These are expressed through the regulatory activities of the *endocrine system*. Many glands secrete hormones or chemicals that stimulate body reactions and turn on or turn off various growth processes.

Study aids

Review Questions

1. Which of the following are primary and which are secondary sex characteristics?
   a. Ovaries
   b. Breasts
   c. Axillary hair
   d. Penis
   e. Voice change
   f. Fallopian tubes

2. According to Anna Freud, what happens to the *libido* at puberty?

3. If an eight-year-old white girl were 50 inches tall and weighed 57 pounds, would she be taller or heavier than most other eight-year-old white girls? How would she compare in height and weight to an eight-year-old black girl in the same percentile group?

4. What is *skeletal age?*

5.   How have patterns of childhood disease changed since the 1930s?

6.   Contrast the Samoan approach to adolescence to Albert Bandura's characterization of the American approach.

7.   Describe the *secular trend* that supports the contention that nutrition is related to size.

Questions Pondered by Psychologists and Physiologists

There is evidence that children are maturing faster physiologically. Is there a related change in the pace of intellectual and emotional growth? If not, what implications does this have for social behavior?

Of Policy Matters and Public Interest

How should schools and families handle increasingly early sexual maturation in a society that is relatively permissive sexually? Some argue that contraceptives should be available to girls and boys of 12 and 13, citing statistics concerning pregnancies and venereal disease in the early teen group. They contend that it is impossible to be sure young adolescents will not engage in sexual activity and that they should know how to lessen the risks involved. Others feel that this is too young an age to bear the burden of decision making about sex and that availability of contraceptives implies a tacit social consent to that activity. Another group suggests that contraceptives be available upon request but that their presence not be advertised to the group at large.

26

Adolescent Personality Development

Student objectives

1. Identify some of the basic conflicts of adolescence: sexual identity, struggle for independence, and establishment of identity.

2. Discuss the significance of the need for reasonable and changing external limits of controls in this period.

3. Discuss the role of rebellion and need for acceptance and belonging in adolescence.

Assignments for this unit

1. Read pages 571–574 and pages 579–627 in *A Child's World*.

2. View Program 26, "Adolescent Personality Development."

3. Read Overview 26 and review the study aids in this book. This overview does not cover any of the material in the text assignment.

Adolescence is characterized by a spurt in physical growth, strong and unsteady emotional states, intellectual changes, and the development of new social relationships. So much is happening that cause-and-effect relations are hard or impossible to disentangle, but experimental evidence and common sense both point to physiological influences as a primary determinant of this period.

The physical changes in adolescence are so far-reaching in their effects that the physical factors affecting personality development and the theories concerning them were included in the physical development chapter in *A Child's World*. We will recapitulate them in this overview, however, for we feel the need to complete the continuum of the personality development theories that we have followed throughout the course.

Psychoanalytic Theory and Adolescent Personality Change

The period of adolescence represents the final stage in Freud's theory of psychic development. It is called the *genital stage* because it is characterized by powerful sexual drives that seek fulfillment in sexual intercourse. The physiological changes that bring about sexual maturity cause a universal concern in adolescents with sexual activity: violent "crushes" on members of the opposite sex, masturbation, sexual fantasies, dating, courtship, and intercourse.

This interest in sexual activity that follows the *latent period* of the middle years is a result of two interacting factors: (1) the physiological activation of genital functioning; and (2) a weakening of the suppressive forces that have kept sexual energy under control. According to psychoanalytic theory, the latter occurs because adolescents transfer the love they have for the opposite-sex parent to an opposite-sex peer. Although the superego still prohibits incest, it is not so powerful in blocking other types of sexual expression, and sexual interest in a friend is not accompanied by so much guilt as sexual interest in a parent.

Healthy personality development requires that tensions be released. Freud was the first to point out this fact and in nineteenth-century Vienna his views on sexuality were revolutionary. Today sex does not appear to play as major a role in neurotic behavior, probably because of changes in social attitudes. Even in Freud's time, however, there was considerable disagreement concerning his emphasis on sexual energy as the driving force of human activity. Josef Adler, a close friend of Freud, proposed that a preoccupation with power is a more major influence on human behavior than sex. People seek to exert their will on others and neuroticism and the adolescent crisis can be explained by the frustrations that accompany strong, unfulfilled drives. Sexual satisfaction is necessary, but its attainment can be seen as a power game. Although Adler's views were not as widely accepted as Freud's, they do point to an important omission in the psychoanalytic doctrine—the social perspective which emphasizes the importance of social drives, such as the need for power or achievement.

The Psychosocial Perspective

Freud was concerned with psychodynamics. He explored the interplay of mental processes and sexual energy. Erik Erikson, on the other hand, brings out observable social aspects of behavior. In each stage he asks what a child does and feels. Erikson's redescription of Freud's genital stage as a period of identity versus role confusion nicely selects the crucial social preoccupation that may replace sexual feelings in today's adolescent. A real need for psychological stability and identity exists under the emotional currents generated by biological processes of sexual maturation. In Freud's day the rigid Viennese social order solved the problem of social identity while it restricted sexual expression. It may be that the situation is reversed today—sexual expression has become more possible and more open while social identity has become the overriding concern.

Adolescents are often shy, insecure, and in need of reassurance. They develop strong attachments to "superior" adults or to people who seem to demonstrate confidence and competence. There is a tendency for social perceptions of people and situations to be oversimplified, presumably because of lack of experience. It is a time of religious conversions, political extremism, and in many cases, of great conformity to the social norms of the peer group.

Adolescence is also a time of risk-taking behavior and rebellion. Teen-agers attempt to test their limits, both physical and social. This is a period of athletic excellence, of overstepping social boundaries, of dangerous experiments and exploits. Adolescents may feel protected through the *personal fable* or feeling that bad things "can't happen to me" (Unit 27). They may enter situations in which they don't yet know social rules. Having found that their parents are not omnipotent, they may also be saying, "If you are not all powerful, your rules aren't either," and they proceed to break those rules. Insofar as teen-agers are exploring the limits of their reality, this kind of behavior serves a real purpose. Automobile insurance rates, at their peak during this period, testify to the more dangerous aspects of this kind of testing.

Erikson sees all these reactions as arising from the need to form a social identity. Adolescents need to know "who" they are. They must make decisions of their own at this point, some of which will affect the rest of their lives. Independence from the family is stressed as the norm by our culture, and both parents and children conform so that by the late teens, adolescents are in charge of most of their life. Their problem, however, is often in deciding what to be and do when they are not sure about who they are.

Under these circumstances it is no wonder that the attitudes and values of friends are powerful influences. Friends are closest to understanding an adolescent's particular problem; he can, in turn, identify with them rather than with his parents who represent the "older generation." At the same time the adolescent is still in need of some kinds of reasonable checks as he moves further toward building his own internal controls. Of course, the amount and kind vary tremendously with age and the individual personality. Both family and society build restraints in ways that shall be explored further in Unit 28.

Study aids

Review Questions

1. What are some of the factors that lead a teen-ager to associate with a particular group of friends? Can you identify ways in which adolescent peer relations differ from those of the middle years?

2. How have attitudes toward sex changed over the last two decades? What qualifications must be added to the statistical data we have?

3. Compare Freud's and Erikson's interpretations of "adolescence."

4. Why might the adolescent years be a period of rebellion? Answer in terms of
   a. Psychoanalytic theory
   b. Your own perspective

5. In what ways might a teen-ager's evaluation of him- or herself vary according to socioeconomic background?

6. What is "generational chauvinism"? Can you relate examples from your own experience?

Questions Pondered by
Psychologists and Sociologists

To what extent is adolescence a cultural rather than a physical phenomenon? What is the dynamic behind changing social attitudes toward sex?

Of Policy Matters and Public Interest

Should juveniles committing legal offenses be treated differently from adults? Children and teen-agers, usually those under 18 years of age, currently receive differential treatment for most offenses. In some cases they are penalized more heavily, in others, much more lightly. Some argue that this is necessary because of their age and inexperience. Others argue that they both deserve equality under the law and should have exactly the same rights and safeguards as adults. These people are not usually the same ones who argue that teen-agers who commit rape, homicide, or robbery can be just as dangerous to society as adults and deserve the same treatment in these cases as well.

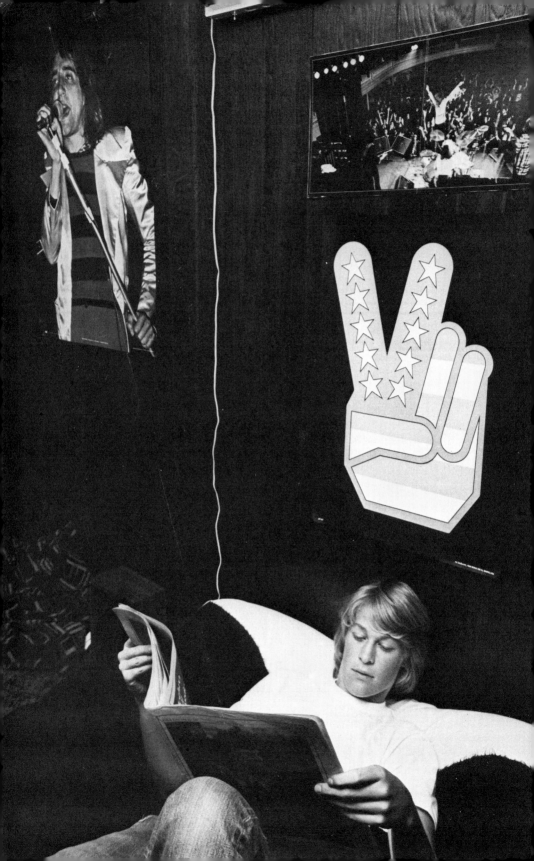

27

Adolescent Mental Development

Student objectives

1. Describe general characteristics of intellectual development in the adolescent period.

2. Discuss Piaget's stage of formal operations.

3. Describe some possible implications (future orientation, change in social attitude, self-concept) of the new cognitive process.

Assignments for this unit

1. Read pages 561–567 in *A Child's World*.

2. View Program 27, "Adolescent Mental Development."

3. Read Overview 27 in this book and review the study aids.

4. Read Floating Objects in this book (optional).

Adolescence, as we have seen, is a period of striking changes in personality and physical development. These changes are so important that they often obscure another important change that occurs in the years between the onset of puberty and adulthood. This change is the development of adult capabilities of thought.

In Piagetian terms, adolescence marks the final stage in mental development, that of *formal operations*. Intellectual development, as measured by IQ tests, levels off after this period unless the tests are affected by unusual experiences in education, specific training, or motivation.

During this period the child-becoming-adult begins to think abstractly, to free himself or herself from a concrete situation, and to formulate hypotheses and philosophies. These new powers of thought are, in most situations, developed by training so that they allow the individual full expression in adulthood. Not all adults, however, attain the capability to think abstractly. The reasons for individual differences in this are obscure, particularly since IQ does not seem to be the predominant factor. Logical thought is probably acquired by social transmission in ways that are not yet understood. Differences in achievement between urban educated individuals and rural uneducated persons as in the experiment by Greenfield described in Unit 19 (page 422 in the text) indicate that cultural factors may be critical for the progression to formal operations.

Intellectual development in adolescence is accompanied by rapid and important physical changes. Emotional upheavals occur as a result of identity changes, ambiguities, and conflicts. The exercise of formal thought helps control and guide the adolescent through crises in this period. *Rationalization* was what Freud called the process through which people try to control unfamiliar feelings and actions. *Projection* and *introjection*, discussed in Unit 26, are other phenomena which are often found in adolescent thought.

The adolescent's challenge to authority occurs in the realm of ideas as well as of standards and conduct, for the struggle to formulate an identity that is consistent with both reason and feeling continues for many years. Teen-agers often embrace or reject ideologies, including religion, on "rational" grounds, but logic fails to explain the totality of the adolescent's ideas. The peer group assumes importance in this area as well as others. It provides a framework for social behavior that can reduce emotional insecurities and uncertainty about what is appropriate. Customs, social forms, and rules are not logical structures, yet they help guide behavior, channel adolescents' energy, and form their thinking about the way they act.

For these and other reasons, absolute consistency in thought is difficult to attain, although the typical adolescent wants it. Extreme, often illogical, positions rather than compromise in both practical and political matters is a recurring result. On a personal level the drive to counter general confusion and dissonance in adolescence can lead to a narrowing of interests. The focus—on music, art, sports, studies—may form the basis for later achievement.

Freud, who saw the individual's entire being as responding to body processes, emphasized the motivational energy of sex throughout this period. Piaget, on the other hand, feels that intellectual development during adolescence as in the other periods is self-regulated and largely independent from other biological systems in the body. In this view, thought does not arise from sexual energy. It emerges because of the inherent tendency for mental *schemata* to become organized and adaptive as well as for them to become increasingly responsive to more aspects of the environment and to the changing nature of the growing individual.

Characteristics of Formal Thought

At the stage of *concrete operations*, children can perform correctly in a number of problem situations that include conserving number, quantity, weight, and volume. A major advance occurs as children move into the stage of *formal operations*. Adolescents can now structure an event in terms of formal or logical relations that exhibit an internal consistency. For instance, they can understand the "law of floating bodies"—that is, if the density (weight divided by volume) of an object is less than the density of water (as in wood), the object will float. If the density of the object (a lead bar) is greater, the object will sink.

This understanding illustrates several new features of thought:

1. The individual is now capable of dealing with two or more complex variables in a situation at the same time, in this case both weight and volume. Piaget has also demonstrated adolescents' ability to handle cases involving three or more attributes of a situation.

2. The variables (weight and volume) are not directly related to what the subject can see, but are specified by schemata developed in the concrete operational stage. Conservation of a liquid quantity, for instance, depends on the realization that if the container is narrow, the liquid level will be higher than if the container is broad. Coordination of these two features (height and width) help determine conservation. Height and width, however, can both be seen. With the advent of formal thought, higher order attributes of situations (such as weight and volume) are taken into account and manipulated formally through the use of mental symbols.

3. The individual comes to perceive statements of fact as statements with logical implications. "The sea is green" is no longer simply a description of a particular scene. It also carries the implication that the sea is not red or any other color but green. This represents the *negation* operation or the adolescents' ability to imagine the situation with something not there or something not done.

The ability to explore different possibilities frees thought from dependence on the immediate present. The accompanying ability to apply logic to a situation means that the adolescents can use thought in the solution of problems to a much greater extent than can children in the period of *concrete operations*. Adolescents can deal with *interpropositional*

thought as well as *intrapropositional* thought. They ca[n]
two statements—(1) Corn needs lots of sun and does[n't do]
well if the weather is cold. (2) The corn harvest was goo[d this]
summer—and make conclusions about the weather. U[nlike]
children in the stage of concrete operations, who would [only]
examine the validity of a single statement, adolescents h[ave]
become able to deal with more than one proposition and determine their logical consistency.

4. The ability to perceive and extend logical implications allows the formulation of systems of thought—
ideologies, theories, philosophies—on the basis of a few, apparently plausible, assumptions. The individual in the stage
of formal operations is theoretically able not only to understand the law of floating bodies but to formulate it from experimentation.

5. The preoccupation with their own thoughts tends to
produce a new form of egocentrism in adolescents. Tom's
opinions appear absolute and those who disagree (often parents) seem stupid or perverse. In most cases continued social
interactions that set his views and arguments against those of
the people around him will modify his belief in his powers of
thought.

In short, adolescents who have attained formal thought
can analyze a situation into its relevant variables, think of all
the possible relations among the variables instead of only
those presented, and imagine situations that are not likely to
be possible. In more concrete terms, if Tom's car breaks
down, he can analyze his problem into components, (fix it
here or walk to a phone), test for the interactions between
these (no tools, but on the other hand, I'm two miles from the
nearest phone), and come up with a solution.

Some of the social and personal effects of the changes
in thought associated with adolescence have been mentioned
both here and in the text. Adolescent *egocentrism*, or concern
with one's own point of view, leads, in many cases, to the
feeling that everyone else is just as concerned with your point
of view as you are. The self assumes a different kind of importance that is both great and vulnerable.

As the ability to think in formal operational terms increases, however, it is often accompanied by more self-confidence. Anxiety diminishes as the adolescent is more in command of situations. The ability to conduct formal operations and free oneself from the immediate present also implies a greater ability to think about the future and past in real terms rather than fantasy. Adolescents can place themselves in time. They can think about the historical forces that shaped them and about the world that they will inhabit when they are ready to find a job. As they become less involved with themselves, they are better able to be concerned about the rest of the world. This type of thinking may be spotty and uncertain as yet; one of the things that may mark the difference between adolescence and adulthood will be the adolescent's growing ability to integrate various fields of interest and thought into a unified whole.

Study aids

Review Questions

1.   What material in this unit might explain why chemistry is normally taught in high school rather than in the fifth or sixth grade?

2.   "Every time I come in the class everyone stares at me." What is this type of belief called in adolescents? How is *personal fable* related to it?

3.   What are the basic reasons for development of thought during adolescence according to:
     a. Piaget
     b. Freud

4.   What conclusion would you draw from the following statements:
     John's party will be called off if there is no holiday.
     John's party lasted until midnight.

a. Was your conclusion based on *interpropositional* thought or *intrapropositional* thought?

b. Would a child in the stage of *concrete operations* draw the same conclusion?

c. Why or why not?

5. Describe two changes in self-concept that might be linked to an adolescent's entering the stage of formal operations.

Questions Pondered by Psychologists

Increasing evidence indicates that only a very few people truly enter Piaget's stage of formal operations. Why is this so? Could education or training help more people move into this stage? Is it necessary for people to enter this stage in order to lead happy, productive lives?

Floating Objects

Children in the conservation stage produce surprisingly naive and confused answers when they are asked about demonstrations of floating and sinking objects. Their progression to a stage (formal operations) in which they are able to understand the law of floating objects has been divided by Piaget into several substages, which illustrate changes in the characteristics of thought.

1. Up to seven or eight years of age, children are satisfied with several contradictory explanations. One object may float because it is light and small, another because it is heavy and large. The children attempt to classify objects into floating and nonfloating but cannot relate specific size, weight of objects, and their own knowledge of the materials that float or sink.

2. The concept of weight relative to water (the object should not weigh more than the water) is grasped at a con-

crete level, but children do not relate the object's weight to an equal amount of water (the amount it displaces) but to the total amount in the container.

3. At about twelve years of age, correct answers appear, accompanied by formal thought.

Why do children who have attained concrete operational schemata of weight and volume fail to relate them in order to discover the law of floating bodies? Because it requires both a simultaneous coordination of two variables (weight and volume) and a formal mathematical operation (division) to solve the formula originated by Archimedes:

$$D \text{ (density)} = \frac{W \text{ (weight)}}{V \text{ (volume)}}$$

An object will float if its weight is less than the weight of the water it displaces.

Only at around twelve years of age is a child capable of acquiring this relation and manipulating it formally. The symbols used in this manipulation are independent of specific content—they can be used with any object or any kind of liquid. Children who have mastered it have arrived at a point where they will be able to think in terms of formal operations.

28

Children in Families

Student objectives

1. Recognize the variety of changing family forms in our society.

2. Discuss positive and negative effects of various family forms on child development.

3. Discuss some of the ways the family influences children's development.

4. Discuss possible effects of family breakdown on children.

Assignments for this unit

1. Read pages 517–532 and 574–579, and review pages 266–270 in *A Child's World*.

2. View Program 28, "Children in Families."

3. Read Overview 28 and review the study aids in this book.

ꣻ Father departs for the office each morning leaving Mother with the two or three children. She uses the second car to drop them at school then runs errands. While Father works downtown, she does the multitude of chores that keep their almost-paid-for suburban home a pleasant base for this typical American family.

This ideal American family is almost outdated, and there may never have been a time when it was more than stereotype. A majority of Americans today would simply not fit into the kind of picture drawn above, although they might strive for it as an ideal, or hold it as a picture of how "normal" people live.

Family sizes have been changing in the United States for some time. In 1790 the first census of the new Republic found that 36% of the household consisted of seven or more persons—a husband and wife with five or more children. In 1940 only 9% of the families fell into this category, and by 1975 the number was 3.5%. The average size of a household in the United States in 1940 was down to 3.77, and this figure had dropped to 2.94 in 1975.

These statistics only represent averages, however, and thus tend to obscure the variety of change related to family life. Women are bearing fewer children, and, at the same time, there is a greater proportion of single-parent families. Because of a greater amount of latitude to divorce or separate, the proportion of children under 18 who were living with only their mothers doubled from 8% in 1960 to 16% in 1975. About one-third of all first marriages currently performed are expected to end in divorce. A high proportion of the young parents this affects will remarry, but still 25% of the nations' children have experienced the disruption associated with dissolution, separation, the death of a parent, or being born out of wedlock, and this proportion may be increasing. For example, it is now estimated that nearly half of all first marriages in California end in divorce.

The reasons behind the apparent decline in commitment to the traditional institution of the family are complex.

CHILDREN IN FAMILIES

Any attempt at an adequate analysis would have to draw on a number of disciplines: sociology, economics, social history, psychology, and so on. In this course our concern is primarily the impact of family structure on the developing child. The influences that determine these structures are not strictly relevant, and although we find them interesting and important, we cannot attempt to deal with them here.

It is, however, necessary to recognize and discuss several family forms in an analysis of contemporary society. The traditional *nuclear* family consists of the two parents and their children. Many scholars consider this unit to be *the* family. There is great controversy over how prevalent the nuclear family is or has been, but recent work in such diverse fields as sixteenth- and seventeenth-century England and modern India indicates that it may be even more widespread than has been assumed.

A household with a single parent (and children) is often called a *fractionated* family, although this term should be used with caution as it implies a prior negative judgment. Implicit assumptions about harm to children because of nontraditional family forms should, in this case and others, be based on objective research rather than value judgements or a resort to stereotypes. There is some indication that such families do seem to have adverse effects on children, but it may be based on incomplete data. Absence of a father makes it difficult for boys to assume a male role, whereas girls may find it hard to relate to men later in life. The absence of a mother—less common but an increasing phenomenon—has not been studied as widely but may be related to high levels of anxiety among the children. As a general rule, however, a proper understanding of the effects of family structure on children is impossible without some knowledge of the wider social context. Some difficulties may be overcome, for instance, through extrafamilial relationships such as friends or Big Brother and Big Sister organizations.

Many societies in the world have a wider kinship group—the *extended family*—as their ideal. In both India and China a family often includes grandparents, unmarried relatives, or even great grandparents, two to three married brothers, and their children all living under the same roof. Certain advantages are obvious—housework, babysitting, and child care is shared, as are expenses; the grandparents remain intimately involved with their children and are taken care of as they age; the presence of other adults provides a variety of caretakers and much adult attention for the children; the continuity of the culture is maintained by the active participation of three and sometimes four generations in the daily business of living.

The chief disadvantage, perhaps, of this extended family is the conformity to rules, duties, and obligations the continuous interaction of so many people often demands. Social roles are rigidly defined for both men and women. The authority traditionally wielded by the oldest members may discourage innovation and creativity among the younger generations. It may not be a coincidence that two of the most stable (or rigid) social systems in the world idealize the extended family. On the other hand, some researchers contend that the association of authority with age is only an ideal in these societies. Real authority, they contend, varies with personality and economic power.

In the United States a kind of "extended family" gained acceptance through the communal movement. A major motivation was to develop a close-knit group, possibly one that was economically self-sufficient, whose members lived in close proximity and shared activities, including childcare. In the 1960s dissatisfaction with the traditional American ideal, rebellion against the alienating effects of many jobs, and revulsion against the Vietnam war led a large number of young people to "drop out" and join communes or other forms of alternative life styles. Today communes do not seem to be increasing, but enough exist to provide some indication of how children adjust to this type of nontraditional situation.

Reports suggest that few communities still possess all their founding members. Some members leave for periods and return, others leave permanently, and new people join. The effect this has on children depends to a large extent on the form child-rearing takes in the community. If children are not able to form stable bonds or attachment bonds to one or two adults who remain with them (such as a mother), the effects of this transience can be devastating. If, as in most cases, there is a primary relationship with the parents, children seem to take to a multiple-caretaker situation fairly well.

Much more has been written about the *kibbutzim* (people who live on a *kibbutz*) in Israel. A kibbutz is a type of communal living arrangement, based explicitly on socialistic principles. All children and adults are to contribute to the group and receive equal benefits. Children are generally brought up in communal houses but are visited by their parents regularly. The general emphasis is on the *collective* activities of children in work and play.

Kibbutz children do maintain the primary relationship but not nearly to the same extent in terms of time spent as children in nuclear families. Many contend that the kibbutz children suffer no ill effects from this style of upbringing; multiple caretakers effectively supplement and substitute for the relationship of a child with its mother. Others have found evidence that kibbutz-born adults have a difficult time forming relationships in any nonkibbutz-style group. There have been other indications that the importance of the peer group and the kibbutz culture takes precedence over individual opinions to a greater extent than they would in American children, but the implications this holds are unclear.

A serious feature of family life in the United States is the high incidence of child abuse. Statistics indicate that child abuse is growing, but it is not clear whether there are more child abusers or whether the reporting of child-abuse incidents is better. Children are battered and often severely or permanently injured by their parents who, otherwise, seem to be normal people. Indeed, this "normality" is one of the most disturbing aspects of child abuse. There seems to be some correlation of low socioeconomic status and lack of education with child abuse, but it may also be true that such incidents in these groups are more often taken to the authorities. Probably more certain is the high incidence of parents who suffered abuse in their own childhoods and who are themselves child abusers.

Is child abuse one of the results of a breakdown in family functions in the United States? This is a difficult proposition to answer. First is the question of whether child abuse is a result of a breakdown of family functions. The answer is probably that abuse can be related to such a breakdown in that a disruption in family relations increases stressful situations for the parents or parent. The second is whether there is, generally, a breakdown in family function.

Traditionally parents perform a wide variety of functions for their children. They protect them and provide them with physical necessities. They also provide emotional support and warmth. They are the mediators for the culture; they are models and provide instruction in social values, norms, and role expectations.

Some have contended that there has been a decrease in parental participation in the third area of responsibility. Television and the high value placed upon peer group interaction provide two very different, but very strong, lines of communication concerning norms and values. The schools are often asked to assume important roles in such diverse areas as teaching children about sex, teaching them how to do household tasks (cooking, sewing, woodwork, gardening), and instructing them in general social behavior. A decrease in time spent with grandparents is paralleled by a decrease in the kind of knowledge about heritage and culture that is gained from them. On the other hand, there is evidence that the first two types of support are provided to an extent that is greater than it has been through most periods of history. Certainly it is true that modern Americans spend a great deal of time thinking, talking, and worrying about the development of their children.

Study aids

Review Questions

1. What parallels might there be between high-income families in the United States and extended families in India?

2. Complete the following table:

Type of family	Members	Possible effects on children
Commune		
Fractionated		
Nuclear		
Kibbutz		
Extended		

3. What problems might there be in interpreting statistics concerning child abuse?

4. Although polygamy is not practied by many Muslims for legal or economic reasons, they are allowed four wives by the Koran. Drawing on the material in this unit, compare the possible effects on children with those of *serial monogamy* in the United States.

5. How have American family forms changed since 1776?

6. What functions do families serve? Drawing on your own experience, relate families you know to the performance of these functions.

Of Policy Matters and Public Interest

U.S. law prohibits polygamy or the practice of having more than one wife. Originally aimed at Mormons in the nineteenth century, this law may technically be applicable to some communes. It is also in our immigration laws—individuals practicing or even advocating polygamy are ineligible for immigration visas. Proponents of this ruling have stood squarely on moral and religious principles. Opponents argue that it violates freedom of religion—since many nontraditional arrangements are religiously based—and freedom of speech.

29

Adolescence to Adulthood

Student objectives

1. Identify the ultimate outcomes of the resolution of the adolescent phase, such as development of sense of self, internalization of values, adaptive self-regulation of behavior, and development of adult sexuality.

2. Discuss the factors that shorten or lengthen adolescence and illustrate their effect on emotional development.

3. Define the mature adult according to various theoretical perspectives.

4. Identify the key social and legal factors that separate adolescence and adulthood.

Assignments for this unit

1. View Program 29, "Adolescence to Adulthood." Because there is no text assignment for this unit, the material in this program will be of particular importance to you.

2. Read Overview 29 in this book and review the study aids.

Overview 29

The introduction to this course dealt with the changing concept of childhood over the centuries. As in any social history, it is difficult to reconstruct what people in less articulate ages thought and felt; but generally, childhood was not viewed in the same manner as it is today. Children, after they had learned to walk and talk, were not thought of as qualitatively different than adults. The differences that were recognized were in size, skill, and physical development.

Those children were often accepted into adult society at what we feel is an early age. For example, boys of 14 were inducted into the army as full-fledged soldiers. The age of marriage has been the subject of greater controversy, but in some areas was clearly lower than it is today. Given the much lower levels of life expectancy that prevailed until recently, perhaps it should not be too surprising that adulthood was hastened.

While the concept of childhood as a different and special period grew throughout the nineteenth century, it was paralleled by the growth of another "stage" of development, that between childhood and adulthood. It was not the first time this intermediate stage had existed in history, but it probably was not as widespread as it has now become. It has certainly been a major factor in the shape of our society.

The concept of this intermediate stage—*adolescence*—was articulated by an early American psychologist, G. Stanley Hall, who wrote his first book on adolescent psychology. His volume, *Adolescence: Its Psychology and Its Relations to Physiology, Anthropology, Sociology, Sex, Crime, Religion, and Education* covered—as one can see from the title—many of the areas that are concerned with adolescence today.

Becoming an Adult

The age at which "adulthood" is conferred by society generally has two dimensions. One involves the *status definition*—the legal ages for voting and drinking, for example—and marks points at which our society formally confers aspects of adult status on adolescents. The other is the *functional* definition. This is usually expressed in relationships and expectations of an individual. It is less clear-cut in our society than the status definition but generally involves location of residence (with parents or independent), occupation, and marital status.

Today the American teen-ager is an obtrusive phenomenon with a particular social identity. This identity is not rigidly synchronized with the biological clock for all adolescents as many reach puberty well before or after the official age of entrance into the teens. Neither is it clearly defined socially. It represents the period spanning the last three or four years of school and the early years of college. Being a teen-ager or adolescent can end at 17 with marriage or a job, but the withholding of adult status can also last, in many ways, well into the 20s.

Where do we draw the line between the adolescent and the adult? In law the problem is solved by an age limit but, as we pointed out, even this is piecemeal. There are varying age limits relating to different activities, and not all states agree on what these should be. Minors (or nonadults) may be individuals under 21 for some purposes and children under 13 for others. The rights to sue and be sued, to vote, to drink alcoholic beverages, to own and drive a car, to marry, or to join the military, all of which are defined in terms of age, vary from state to state.

Movement into a functional role as an adult can be even more confused. Many graduate students in their late 20s feel they aren't accorded full adult status because they don't have regular jobs. Boys who have dropped out of high school and marry at 18 are catapulted into that same adult status.

From the point of view of the psychologist, adolescence and adulthood should refer to states of mind rather than social or economic conditions. These other factors, however, are clearly reflected in the individual's view of himself and in his outlook on life and can hardly be separated from them. One tacit assumption is that an adult is "mature," aware of social responsibilities and willing to be guided by them. Unfortunately the concept of maturity is too vague to be measured easily and is not the same as reaching voting age, having a job, or owning a home and a car.

The earlier chapters on adolescent mental and personality development have described the challenges and problems in the vague but real period of transition between childhood and adulthood. In this chapter, the focus will be on growth to adulthood. We must remember, though, that these later influences are largely emotional, social, and economic. The physical changes of puberty have stabilized during the teens and, sexual interests have begun to influence personality and motivation profoundly. Intellectual growth, in terms of IQ, seems to plateau at about midadolescence. Changes in mental ability and attainment in adulthood are supposedly a result of increases in knowledge, expertise, and experience. It is in adolescence that the stage is set for the growth into adulthood and—if it ever comes—maturity.

Theories of Maturation into Adulthood

In many traditional cultures the transition to adulthood was fairly abrupt and often marked by a ritual. The Jewish Bar Mitzvah is an example; even the Anglican Confirmation may be interpreted in this manner. Today youths may maintain life styles for many years into their 20s with little outward change to characterize adulthood; yet psychological and social growth usually does not stand still.

We do not really understand the socialization process of adolescents as yet. Scientists have to be content with close observations, survey data, and their intuition from which they devise theories of the adult personality and its vicissitudes.

Scientists are unlikely to unravel cultural pressures and biological forces and their interaction on the developing personality in the near future.

Freud and his followers have provided the richest insights into the adult mind. Freud did not provide another stage after the *genital* period of adolescence, but he did define mature adults as those who had mastered their early Oedipal conflicts (See Unit 15). The person is ready to assume the role of a parent and to relate to others.

Erik Erikson takes Freud's material and elaborates on it, but he deemphasizes sex as a basic determining drive. Erikson stresses the importance of social influences on the individual. Erikson's adolescent is immersed in a crisis of *identity* (Chapter 26). Young people are typically caught up in the turmoil of physical changes and new social expectations. A stable sense of who they are arises out of social experience and experimentation. However, for some the early instability is too great to acquire the necessary social skills; thus social isolation can result when adolescents leave the family to make a life of their own.

This isolation is part of Erikson's characterization of early adulthood. Adolescence is a time of *identity versus role confusion,* and the subsequent years are a period of *intimacy versus isolation*. This theory proposes that adults' ability to relate to others in a meaningful way depends on the resolution of earlier identity conflict. A stable, well-delineated sense of self allows greater flexibility and confidence in social relations. The individual feels free to give of himself to others without fear of rejection, enjoys social relations, and forms close friendships. From these attainments the desire for marriage and its associated emotional responsibilities arise. Later, in the 30s perhaps, more focused intellectual skills tend to allow creativity and achievement in one's chosen fields of interest. This later period Erikson calls *generativity versus stagnation*. A positive response at this stage will ensure a favorable position in the final stage, *ego-integrity versus despair* (see Table 6-1 *A Child's World,* p. 221).

Erikson's descriptive categories make intuitive sense, especially as they include the whole life cycle. Psychologists have generally conceived of development as culminating with adolescence rather than continuing through life. The reason for this lack of interest in adult development may lie in its greater familiarity and apparent "normality." Changes, both psychological and social, may seem slow and undramatic. Responses to environmental changes are assumed to be characteristic of the "stable," mature personality which was formed during childhood. Both the early philosophers and Freud emphasized the importance of childhood for understanding the adult, but the adult's behavior was not thought to involve new departures or surprises. This assumption is worth reexamining, perhaps, especially in view of the greater options which our society now allows adults.

Carl Rogers provides an alternative view of adulthood to that of the psychoanalysts. For Rogers individuals seek to express their uniqueness and relate to others. Rogers has not developed a systematic theory of development, but his "humanistic psychology" emphasizes the "person" and his or her continuing growth. A mature person, in Roger's view, is in touch with his own values and desires and accepts them.

What *is* an adult? Perhaps we all spend most of our lives seeking greater maturity, certainly there is no one psychological definition. Douglas H. Heath has formulated a model of the maturing person that we have found useful. It is summarized by Rolf Muuss as follows:

> The mature individual has intellectual competencies that allow him to conceptualize and to symbolize his own experiences more effectively. He can bring into consciousness the motives underlying his behavior and reflect upon them . . . becoming mature implies becoming more allocentric [other-centered] in social and interpersonal relationships . . . without becoming conforming. The mature person appreciates social relationships more, and he can communicate more effectively. . . . Growth toward maturity implies better integration of various aspects of one's personality. Thought processes, philosophy of life, self-image, sex, and love become more integrated with one another, establishing a more unified identity. To achieve such integration in the

face of social changes and increasing social complexity implies openness to experiences, flexibility in problem solving, and a continuous, active involvement in one's own development . . . self-image, value systems, interests, and interpersonal relationships become more stable with increased maturity. Adolescent infatuations are replaced by more permanent and more altruistic love. In all of these areas of life greater resistance to emotional disruptions and social disturbances has developed . . . the mature adult has acquired autonomy in thought processes, values, and decision making. He is less influenced by parents, peers, friends, and advertisements. In spite of social pressure, verbal persuasion, or the lure of the ad he will maintain his own independence, autonomy, and conviction. He reasons more objectively and can suspend decisions and judgement until sufficient relevant information has been accumulated.*

Study aids

Review Questions

1. Divide the following into attributes of (1) a functional definition of adulthood or (2) a status definition of adulthood.
   a. Right to vote
   b. Steady job
   c. Ownership of a home
   d. Driver's license
   e. Minimum age of marriage
   f. Parenthood

2. How is the *identity versus role confusion* that marks Erikson's stage of adolescence related to the *intimacy versus isolation* of early adulthood?

*Rolf E. Muuss, ed., *Adolescent Behavior and Society* (New York: Random House, 1975), p. 614.

3. How does Douglas Heath characterize the mature adult? Does this characterization coincide with your personal definition of maturity?

4. How has the concept of adolescence changed over the years?

5. What factors have caused a lengthening of the preadult or adolescent period in our society?

Questions Pondered by Psychologists

Is it possible—or wise—to identify clear boundaries marking adolescence and adulthood? How much predictable change is present in the adult mind and personality as the individual ages?

Of Policy Matters and Public Interest

Should compulsory education be abolished after the eighth grade? Proponents argue that it should, adding that we retain many youngsters in school who do not wish to be there and who create difficult situations for those students who do wish to learn. They suggest that each individual be given the opportunity to receive up to 14 years of education at any point during their lifetimes rather than restricting education to a certain period that we now call adolescence. Opponents point out that the demand for unskilled labor is low and contend that young teen-agers are not yet mature enough to move into adult roles.

30

Conclusion

In this telecourse we have examined human life in a developmental sequence, from its origin in the fertilized ovum to the adult person. This miracle of growth has always been a source of fascination, although science has only recently paid attention to the possible mechanisms involved in development. Perhaps it is the "new" status of this science that leads to these final reflections upon child development and the theories we have reviewed in explaining it.

Throughout this course several theoretical approaches to child development have been presented. At present there is no unified view concerning them, and each of us has his or her own biases. We hope that, by now, you will not assume that any psychologist or social scientist offers definitive answers about child development—but neither should it be assumed that all theories are equally wrong.

It is hard to judge theories in psychology by the harsh standards of "truth." No theory is absolutely true, even in physics. A theory may clarify a set of phenomena without helping to solve the larger problems. Another theory may not completely explain any details yet can provide a good overall orientation. Both types of theory are needed and exist in psychology; what has not yet been done is to find one theory that does all things.

An influential theory in the United States for most of this century (almost the whole lifetime of psychology in this country) has been the behaviorist, or S–R approach. Inspired by the Russian physiologist, Pavlov, many U.S. scientists believed all human behavior could be reduced to chains of simple reflexes or responses triggered automatically by stimuli. The S–R theory, centered not on humans but on a simpler animal, the rat, was assiduously pursued for decades. It is reasonable to say that its high promise has not been fulfilled.

More recently Piaget's theory has dominated developmental psychology. Piaget's emphasis on assimilation and accommodation points to the fact that organisms *interact* with their environment and that both together form the *system* which is the appropriate unit of study. Behavior is not simply the result of emitting signals generated internally, nor is it responses to signals (stimuli) impinging from outside. Behavior is the result of both internal and external influences.

The nature of the external influence can be thought of as "information," whereas its processing goes on inside the brain. This is the *information-processing* view of cognition, of which we have said little. It is a relatively recent approach, based on the computer analogy. One important point it makes is that we have to understand the structure of informational input, the structure of the processing mechanisms, and the nature of the processing before behavior can be explained scientifically.

The view of the Gibsons, whom we have mentioned occasionally, is based on information and its detection but does not delve into the mechanisms of how this occurs. They do not propose psychological constructs to "explain" the workings of the brain or mind; rather they explain changes in actions through changes in perception. Perhaps their approach emphasizes a more modest orientation to facts, a kind of descriptive psychology of humans in their ecological setting. This approach has strong theoretical implications which some of you may encounter if you continue in psychology.

In some areas, particularly in the chapters on personality development, we have studied two other bodies of theory. These are the social learning theory propounded by Bandura and his associates and the approach to psychology taken by Freud and his followers. The Freudian influence, amplified and extended by Erik Erikson, has been a pervasive one, often coexisting with one or another of the theories of mental development mentioned above.

Will any of these attempts to "explain" human behavior and development prove adequate? Perhaps the very notion of people trying to explain their own workings is ludicrous. Progress has been made, however, and even if only a small part of the human experience is eventually systematized, it will be an achievement for science.

Furthermore, we should not neglect the important social consequences of child development. The teaching of this science helps foster an objective viewpoint. This need not mean—should not mean—treating ourselves as machines or computers, but rather looking on the human condition as something within our power to modify and improve. There has been a record of success in treating mental illness, for example, that has altered people's view of "madness" from the notion of possession by demons to that of a state of chemical imbalance. The possible benefits of controlling and modifying ourselves are obvious and so are the dangers if success in this task dehumanizes. Just as doctors are told to treat the patient and not the disease, so psychologists in trying to understand child development must study the child as a biological *and* social being rather than a bundle of mechanisms. To do otherwise would be bad science and, in our opinion, poor philosophy.

Glossary

Accommodation (in Piaget's theory) The modification of an existing schema by which a person perceives or thinks as a result of new experiences.

Adaptation (in Piaget's theory) Refers to the twin processes of assimilation and accommodation. It involves the creation of new schemata or modification of old schemata through which the child might effectively deal with the immediate environment and is therefore the essence of intelligent behavior.

Adapted Information Information designed to communicate to the listener.

Aggression Behavior aimed at hurting.

Alleles A pair of genes that affect the same trait.

Amniocentesis A process by which amniotic fluid is extracted from the womb, usually during the second or third trimester, in order to determine the presence of birth defects of Rh disease.

Amniotic Sac A fluid-filled membrane that encases the developing infant.

Anal Period In Freud's psychoanalytic theory the second psychosexual stage in which pleasure is focused on the functions of elimination. Begins around 12 to 18 months and ends at about three years of age.

Androgens Hormones associated with the development of masculine characteristics.

Animism The tendency to ascribe life, consciousness, and will to inanimate objects.

Anoxia Lack of oxygen.

Apgar Scale A series of tests measuring the degree to which a newborn is physically and behaviorally normal.

Areolas Pigmented areas that surround the nipples.

Artificialism   The Piagetian stage of development of causal thinking in which children believe familiar adults or other people have created everything that exists.

Assimilation (in Piaget's theory)   Refers to the incorporation of new aspects of stimulation into the existing schemata.

Associative Play   Preschool play in which children play with each other in a common activity but without subordinating their interests to those of the group.

Attachment   Refers to the affectional bond linking persons to each other.

Autosome   All chromosomes except sex chromosomes.

Axillary Hair   Refers to the hair that begins growing under the armpits during adolescence.

Babbling Stage   Period of repetition of meaningless sounds by infants prior to the onset of speech.

Babinski Reflex   An infant reflex in which the toes fan out and the foot twists inward in response to the sole of the foot being stroked.

Babkin Reflex   An infant reflex in which the infant responds to pressing of the palms of the hands with turning of the head and opening the mouth.

Basal Age   Associated with the Stanford-Binet intelligence test, the highest age level at which a child can answer *all* questions for that level.

Behavior Modification (Behavior therapy)   The application of operant learning and other experimental laboratory procedures to change human behavior.

Bell Curve   A graph illustrating the distribution of IQ scores in the population at large.

Blastocyst   In prenatal development a single, spherical layer of cells enclosing a fluid-filled central cavity. It occurs during the germinal stage of pregnancy.

Castration Complex   In Freudian psychology the result of childhood fears of punishment for forbidden sexual desires toward parents of opposite sex. In male children it is manifested by fear of losing genitals. In females, by a fantasy that the penis has been removed from her as a punishment.

Cathexis   The investment of emotional energy in an idea, object, or activity.

Centration (in Piaget's theory)   Refers to the focusing of attention on one aspect of a situation while ignoring all others.

Cephalocaudal Growth   Development beginning at the head then proceeding downward to other parts of the embryo.

Cervix   The opening to the uterus.

Chromosome   The 46 rod-shaped particles in the cell that determine the genetic make-up of an individual.

Chronological Age   One's age calculated from birth, rather than from conception.

Class Inclusion (in Piaget's theory)   Refers to the understanding that a part is a fraction of a whole (e.g., a daisy is part of the "class" flower).

Classification   Ability to sort objects or concepts into categories according to their specific attributes.

Cleft Palate (hairlip)   A condition in which the roof of the mouth fails to fuse completely.

Clinical Method   A method of assessing a child's thinking ability developed by Piaget that allows the child scope for natural behavior in the situation while probed by questions or presented with problems by the investigator.

Cognitive Process   A mental activity such as using language, thinking, reasoning, solving problems, conceptualizing, remembering, imagining, or learning verbal material.

Cognition   *See* Cognitive Process.

Collective Monologue   A situation in which a group of two or more children are talking with no communication achieved or apparently intended.

Compensation   Recognition in the final stage of conservation that a change in one aspect may be accompanied by an opposite change in another.

Concrete Operations (in Piaget's theory)   The stage of mental development, roughly from 7 to 11, in which the child demonstrates a growing ability for mental activities but has not yet achieved full mastery of the use of abstract concepts.

Consanguinity   Describes a close relationship by blood.

Conservation (in Piaget's theory)   The recognition of the permanence of objects under varying conditions. It involves the ability to recognize that two equal quantities of matter remain equal even if rearranged.

Control Group   One or more groups that are not subjected to changes in the independent variable but are otherwise exposed to the same setting and procedures as the experimental groups in an experiment.

Convergent Thinking   The ability to follow accepted patterns of thought and arrive at single correct solutions to particular problems; this ability is generally assessed by intelligence, aptitude, and achievement tests.

Cooperative Play   A type of play activity in which children play in organized groups with differentiated roles and commonly accepted rules.

Correlation   A statistical parameter indicating the degree to which two variables are associated or vary together (e.g., weight and height).

Critical Periods   Those times during development when the organism is most sensitive to the effects of the environment.

Cross-Sectional Study   A one-time study involving a large number of individuals who are compared on the basis of one dependent variable while they differ in known ways with regard to one or more independent variables.

DNA (Deoxyribonucleic Acid)   A molecule which is found in the nuclei of cells and is responsible for the transmission of hereditary characteristics and for building proteins.

Darwinian Reflex   Infant reflex in which a strong fist is made when the palm is stroked.

Decenter (in Piaget's theory)   The ability to take more than one aspect of a situation into account in order to solve a problem.

Deciduous Teeth   The primary teeth of childhood that will eventually fall out and be replaced by permanent ones.

Deductive Reasoning   Reasoning from the general to the particular.

Deferred Imitation   Imitation that does not follow immediately upon the stimulus behavior but follows sometime later based on the mental symbol for it that has been retained in the mind.

Dependent Variable   In an experiment the variable whose state is caused by, and therefore depends on, changes in another variable, called the independent variable.

Developmental Quotient   A figure computed by dividing a child's mental age by its chronological age and then multiplying by 100.

Differentiation   Progression from the general to the specific. In motor development increasing control over more specific movements. In mental or emotional development the increasing ability to discriminate on the basis of more specific apprehension.

Divergent Thinking   Innovative and original thinking that deviates from customary patterns of thinking and results in more than one correct solution to particular problems; this ability is generally assessed by creativity tests.

Dizygotic Twins   *See* Fraternal Twins.

Double Helix   Spiral arrangement of the two strands that comprise the DNA molecule found in chromosomes.

Echolalia   Conscious imitation of the sounds of others.

Ectoderm   In prenatal development the upper layer of the embryonic disc that will develop into the epidermis, nails, hair, teeth, sensory organs, and nervous system, including brain and spinal cord.

Ectomorph   A tall, thin body type.

Ego   One of three parts of personality postulated by Freud. The ego develops to handle transactions with the environment, the id, and superego.

Egocentric   Lack of differentiation between self and the outside world to the extent that the self's thoughts and desires are viewed as central to reality.

Egocentric Speech   Speech not apparently directed to a listener.

Elektra Complex   Female counterpart of the Oedipal complex in which, theoretically, girls sexually desire their fathers and fear their mother's jealousy.

Embryo   In prenatal development a blastocyst that has been fully implanted in the uterine wall.

Embryonic Disc   Thickened cell mass from which the embryo will develop.

Embryonic Stage   Second stage of prenatal development, lasting from the second to the eighth week and characterized by rapid growth and differentiation of major body systems and organs.

Endoderm   In prenatal development the lower layer of the embryonic disc that will later develop into the digestive system, liver, pancreas, salivary glands, and respiratory system.

Endomorph   A round, plump body type.

Equilibrium (in Piaget's theory)   Refers to the balance between assimilation and accommodation.

Erythoblastosis Fetalis (Rh hemolytic disease)   A disease in newborn infants caused by a reaction of the immunity system of the Rh negative mother to the infant's own Rh positive blood.

Estrogens   Feminine sex hormones that influence the development of secondary sex characteristics and regulate the estrus cycle.

Experimental Group   In an experimental situation the group whose responses or characteristics are altered by the independent variable in order to test the validity of an hypothesis.

Extended Family   Social unit consisting of parents, children, and other adults and children (often blood related). Is usually limited to those living in one household.

Factor Analysis   Technique used by statisticians to locate an element common to a wide variety of items.

Fallopian Tube   A slender tube leading from the ovary to the uterus.

Fetal Stage   Final stage of prenatal development, lasting from the eighth week until birth, characterized by rapid growth and by changes in body form.

Figural   Refers to the two-dimensional pattern or form of a scene or object.

Fixed-Pattern Reaction   A set of reactions that is invariant to the presentation of a particular stimulus.

Follicle   A small ovarian sac in which the individual immature egg (ovum) is encased.

Fontanel   Space between the bones in the infant's skull covered by a membrane and commonly known as the "soft spot."

Formal Operations (in Piaget's theory)   A stage of cognitive development occurring around the age of 12 and characterized by the increasing ability to engage in hypothetical reasoning.

Fractionated Family   Refers to a social unit that consists of only a part of the nuclear family. *See also* Nuclear Family.

Fraternal Twins   Refers to the fertilization of two ova released within a close proximity of time and the consequent development of a set of twins, each with its own unique genetic make-up.

Functional Invariants   Attributes the function of which does not change in spite of other changes that may occur.

Gametes   Sex cells: refers to either egg or sperm.

Gene   The basic unit of heredity. By controlling enzymes and protein production, genes direct the development of internal and external bodily structures, duplicating those that are inherited from the parent.

Genotype   Describes the actual genetic composition of a specific trait.

Germinal Stage   The earliest stage of prenatal development, lasting ten days to two weeks from the moment of conception and characterized by rapid cell division with implantation of the organism in the uterine wall.

Gestalt Psychology   The theory in psychology that the psychological experience is based on holistic relations rather than elements of sensation (not to be confused with Gestalt Therapy).

Gonads   The sex glands: ovaries in the female and testes in the male.

Gonatropic Hormones   Sex-appropriate hormones that are released in increasing amounts just prior to puberty.

Habituation   The process by which an individual becomes accustomed to a certain set of stimuli.

Heterozygous   Describes an allelic pair in which the maternal and paternal genes are different.

Hierarchic Integration   The incorporation of individual abilities into larger complex behavior patterns.

Homozygous   A term used to describe an identical pair of genes.

Horizontal Decalage (in Piaget's theory)   A term describing the child's inability to transfer learning about one type of conservation to a different type, although it may be based on an identical principle.

Hydrocortisone   A hormone secreted by the adrenal cortex in times of stress and thought to be linked to the development of the condition known as cleft palate.

Hyperactive Child Syndrome   A complex assortment of behavioral traits that include impulsivity, restlessness, an inability to concentrate, a high level of activity, and emotional lability.

Hypophysis   A fetal endocrine gland thought to signal the beginning of labor.

Hypothetico-Deductive Reasoning   Reasoning process in which an hypothesis is developed and its deductions tested experimentally in order to determine whether or not it is true.

Id   One of three parts of personality in Freud's theory that is the source of all instinctive energy and strives for immediate gratification.

Identical Twins   Refers to twins who develop from the same fertilized egg and have identical sets of genes.

Identification (with other people)   A process by which the characteristics of another person are incorporated into the self.

Idiopathic Hypopituitarism   A type of growth hormone deficiency in which the activity of the pituitary gland is abnormally diminished for no known cause.

Independent Variable   In an experiment the variable that is manipulated and causes changes in another variable (called the dependent variable).

Inductive Reasoning   Reasoning from the particular to the general.

Inhibition   Refers to constraint or halting of either behavioral or physiological ongoing activities.

Intellectualization   To consider the rational content of an issue while ignoring the emotional or psychological significance by using an excessively abstract explanation.

Intelligence Quotient   A way of measuring a person's relative performance on an intelligence test. It indicates how a person compares with others in the same age group.

Interpropositional Thought   The ability to use the content of more than one statement in order to draw appropriate conclusions that are not explicitly stated.

Intrapropositional Thought   The ability to think about the content of one statement.

Introjection   The child's incorporation of aspects of the same-sex parent's personality into its own through identification.

IQ   *See* Intelligence Quotient.

Karyotype   A chart used to demonstrate chromosomal structure.

Kibbutz   A communal living arrangement in Israel in which adults live apart from children during the day with children being raised together by persons trained to care for them and teach them, thus freeing all other adults for productive work.

Lallation   In early language development the repetition of sounds and syllables heard in the environment with imperfect imitation.

Lanugo   Fuzzy prenatal body hair generally lost either at birth or shortly thereafter.

Latency Period   In Freudian theory the stage encompassing the ages of approximately 6 to 12 years and characterized by relative sexual inactivity.

Libido The psychic energy of the id, considered by Freud to be sexual in origin.

Lipids Fats.

Longitudinal Study A study that looks at the same subjects or particular characteristics at several different points in time to examine changes that occur over time.

Matrilinear Families in which the primary identification of lineage is through the mother.

Maturation The emergence of behavior patterns that depend primarily on the development of body and nervous system structures.

Meconium Fetal waste matter formed in the fetal intestinal tract, present at birth and excreted during the first two days after birth.

Meiosis The division of a cell for sexual reproduction in which the resulting cells have only 23 chromosomes.

Menarche The onset of menstruation.

Mental Age Age assigned on the basis of relative performance on an intelligence test. It indicates intellectual ability in terms of an age level.

Mesoderm In the blastocyst in prenatal development the middle layer of cells that develop later than either ectoderm or endoderm and that will eventually differentiate into the dermis, muscles, skeleton, excretory, and circulatory systems.

Mesomorph A muscular body type.

Mitosis The process of division in all cells except those for sexual reproduction during which the chromosomes divide, giving each new cell a full complement of 23 pairs.

Monogamous Refers to the relatively permanent pairing of two individuals on a one-to-one basis.

Monologue In language development it refers to the lone young child's speech that is apparently not directed toward another individual.

Monozygotic Twins *See* Identical Twins.

Moro Reflex An infant reflex in which the body becomes fully extended with back arched in response to a sudden (startling) stimulus.

Mutations Changes in the gene that cause new (generally harmful) characteristics to be produced in the mature organism.

Naturalistic Studies Studies that depend entirely on observation with no attempt to alter behavior or manipulate circumstances.

Negation Operation (in Piaget's theory)   A mental operation that negates an existing state of affairs and that may emerge as an action.

Neonatal Stage   Refers to the first two weeks of life.

Neonate   An infant during the first two weeks after birth.

Norms   Averages of test performance of a large group of people (standardization sample) that allow a test examiner to interpret an individual score.

Nuclear Family   Traditionally a social unit consisting of mother, father, and children. This term is often now used to refer to any exclusive social unit based on this structure consisting of two adults and children.

Object Permanence (in Piaget's theory)   Refers to the realization that an object remains in existence even though it cannot be sensed in any way.

Ocular-Neck Reflex   An infant reflex in which the neck is bent back in an effort to avoid a light flashed in front of the infant's eyes.

Oedipal Complex   In Freudian theory the sexual attachment to the mother by a male child with corresponding hatred, or jealousy, toward the father. (Male equivalent of Elektra complex).

Operant Conditioning (Instrumental Conditioning)   The process of modifying the frequency of an active response that seems to be under voluntary control by modifying the consequences that follow it.

Operational (in Piaget's Theory)   See Concrete Operations.

Oral Period   According to Freud's psychoanalytic theory the first psychosexual stage during which the child's pleasure is focused on the mouth and oral activities such as sucking and eating (ages 0-18 months).

Organization (in Piaget's theory)   The integration of all functions into a consistent and optimal system.

Orienting Reflex (OR)   Reflex in which the organism responds to a stimulus by turning its head physically toward it while stopping all other activity.

Overextension   In language development the inappropriate use of one word to mean several different things.

Overgeneralization   In language development a general statement, rule, or law applied beyond its normal limitations (e.g., application of past tense "-ed" to all attempts to indicate past tense, such as "goed" rather than "went").

Ovulation   The process in which the mature egg is expelled from its follicle to begin its journey toward the uterus.

Ovum (pl. Ova) Female germ cell or egg.

Oxytocin The hormone produced in the pituitary gland and responsible for the mechanism that takes milk through the ducts in the breast to the nipple.

Parallel Activity A type of preschool play in which a child plays in close proximity to other children but does not attempt to influence or modify their activity.

Patellar Reflex Knee-jerk reflex in which the leg moves responsively when the knee is tapped.

Patrilinear Families in which the primary identification of lineage is through the father.

Perception Refers to the apprehension of the environment as the result of the extraction of information from stimulation.

Perinatal Stress Stress conditions experienced around the time of birth.

Phallic Stage In Freudian psychoanalytic theory the third psychosexual stage when pleasure is focused in the genitals and the Elektra or Oedipus complexes arise (ages 3–5).

Phallocentrism The belief (typified by Freud) that the male is the ideal and the norm by which both males and females are judged.

Phenotype Refers to the expression of any particular gene or set of genes.

Phocomelia A congenital defect characterized by abnormally short limbs.

Placenta An organ formed from the blastocyst that conducts nourishment to the developing embryo and the embryo's wastes to the mother.

Placing Reflex An infant reflex in which the infant stiffens its legs when the backs of its feet are resting against the edge of a flat surface.

Polyandry Traditionally, a social arrangement characterized by one female and two or more males in lasting, socially sanctioned sexual relationships that may result in children.

Polygamy Traditionally, a social arrangement characterized by one male and two or more females in lasting, socially sanctioned sexual relationships that may result in children.

Preconcepts (in Piaget's theory) The child's earliest concepts that are active and concrete rather than schematic and abstract.

Prematurity   Refers to the birth of a child before the end of the 37th week of gestation who weighs less than 2500 grams.

Preoperational Stage (in Piaget's theory)   Describes the child between ages two and seven, whose thought is egocentric but who has begun to use symbols. Lack of conservation is a chief characteristic.

Primary Circular Response (in Piaget's theory)   Refers to an infant's actions related to its body (e.g., thumb sucking) usually between birth and four months.

Progesterone   A feminine sex hormone whose primary function is to prepare the uterus for pregnancy.

Projection   A defense mechanism that occurs when people assign their own undesirable characteristics, mistakes, problems, impulses, desires, or thoughts to others in order to reduce their own anxiety at having to recognize these characteristics as their own.

Prolactin   A hormone released from the pituitary gland at the time of delivery of the placenta, which stimulates the production of milk.

Proximodistal   Development beginning in the central part of the body then proceeding to the extremities.

Psychometricians   Those who measure the mental abilities of individuals.

Puberty   That point in physical development at which an individual is sexually mature and able to reproduce.

Pubescence   Stage of rapid physical growth, lasting about two years, marked by maturing sex organs and appearance of secondary sex characteristics.

Reaction-Formation   A defense mechanism that occurs when persons conceal a real motive or emotion from themselves and express the opposite one by attitudes and behavior—presumably to avoid anxiety associated with the real motive or emotion.

Realism (in Piaget's theory)   The confusion of the child's ideas and feelings with objective reality.

Reflex Behaviors   A motor act automatically elicited by stimuli in the environment. A rapid, consistent unlearned response that is not generally subject to voluntary control.

Regression   A defense mechanism characterized by the return to immature modes of behavior used originally at a younger age and elicited by stress or anxiety.

Repression   A defense mechanism that occurs when people exclude anxiety-arousing motives, ideas, conflicts, thoughts, or memories from awareness.

Reversibility (in Piaget's theory)   A term that refers to the awareness that a transaction or event can be reversed to reinstate the original condition.

Rh Factor   Refers to a protein substance in the red blood cells.

Rh Hemolytic Disease   *See* Erythoblastosis Fetalis.

Rooting Reflex   An infant reflex in which the head turns toward a finger or nipple brushed against the cheek, as the mouth opens.

Schema (schemata) (in Piaget's theory)   An organized mental structure that applies to both mental operations and actions. (Piaget distinguishes between a scheme and a schema, but we have not.)

Schizophrenia   A group of psychotic disorders likely to include many of the following patterns: faulty perception, disorganized thinking, emotional distortion, delusions and hallucinations, withdrawal from reality, bizarre behavior, and incomprehensible speech.

Secondary Circular Response (in Piaget's theory)   Stereotyped actions that are directed toward achieving an effect in the environment (e.g., repetitive banging to make a noise).

Secular Trend   A tendency toward change in a particular direction or of a particular type occurring over a long period of time.

Sensorimotor Stage (in Piaget's theory)   The stage of cognitive development that begins at birth and extends to around the age of two. Characterized by learning and thought arising from sensorimotor behavior.

Serial Monogamy   Refers to the social unit formed by two individuals who remain paired until one or both decide to withdraw from the union. At such time each is free to form an exclusive relationship with someone else.

Seriation   Refers to the ability to arrange objects or concepts in a sequence along one or more relevant dimensions.

Siblings   Generically, children related by blood. Used also in reference to children raised together within an exclusive familylike social unit.

Sign   An arbitrary conventional item in a representational system (such as a word in a language) that refers to some aspect of the world.

Signal   An object or event that refers to other objects or events, in the past, present, or future (e.g., a crashing sound signals an accident).

Significate   The referent or whatever is referred to by the signifier.

Signifier   An object, person, or event that comes to mean (or signify) something to the child.

Socialized Speech   Speech activity in which communication is intended.

Sociogram   A representation of closeness and distance and so on of relationships among people in pictorial form.

Sociolinguistics   Study of the interrelationship between language and its social environment.

Sociometric Techniques   Methods through which social acceptance or rejection are measured based on preferences expressed within a social group.

Solitary (Independent) Play   A type of preschool play activity in which the child plays alone, even when other children are nearby, using toys that others are not using and focusing complete attention on its own activity.

Spermatozoon (pl. Spermatozoa)   The male germ cell or sperm.

Spontaneous Abortion   Otherwise known as a miscarriage, refers to the expulsion of a nonviable embryo or fetus from the uterus.

Standardization   A process by which test items are evaluated by administering them to groups of children similar to those for whom they have been ultimately devised.

Standardization Sample   In a test situation this is the representative group first tested that is similar to the group for which the test has originally been designed.

Stanford-Binet   A standardized intelligence test developed for use with children between 2½ and 18 years of age.

Sublimation   A defense mechanism in which anxiety-producing sexual energy is channeled into acceptable activities such as intellectual interests and achievements.

Superego   The last part of Freud's personality structure to be formed. It incorporates parents' and society's values and functions as a "conscience."

Swimming Reflex   An infant reflex in which the infant exhibits involuntary, yet well-coordinated swimming motions if placed stomach down in water.

Symbol   A representation or depiction of an object or event that is in some way similar to the object or event.

Symbolic Function   The ability to represent objects, places, or persons in the mind when they are absent from the present scene.

Telegraphic Speech   In language development speech containing only words that convey meaning and in which articles, prepositions, and word endings are often absent.

Tertiary Circular Response (in Piaget's theory)   Actions that are *intended* to achieve some result on the environment. The child seems intent on attaining the goal.

Testes (sing., Testis)   Also testicles; the male reproductive gland, located in the scrotum, in which spermatozoa are produced.

Thalidomide   A tranquilizer linked to birth defects in infants whose mothers used the drug early in pregnancy.

Time Sampling   Involves repeated recording of a behavior or behaviors during a specified and recurring period of time.

Tonic Neck Reflex   An infant reflex in which the infant who is placed on its back will turn its head to one side or the other and extend the arm and leg on the preferred side while flexing the limbs on the other.

Toxoplasmosis   An infection caused by the parasitic protozoan *taxoplasma gondii*, barely noticed by most people who contract it but causing brain damage, blindness, or even death to the unborn child.

Transductive Reasoning   Reasoning from one thing to another without considering the hierarchical relationships involved.

Transformations (Psychological or Linguistic sense)   Changes in form or structure.

Transitivity   The ability to recognize the relationship between two objects by apprehending common relationship to a third (e.g., Jane is taller than Bill who is taller than Frank).

Trophoblast-Villi   In prenatal development the outer cell layer (ectoderm) of the blastocyst from which threadlike structures emerge (villi) to penetrate the uterine wall thereby allowing the blastocyst to become firmly implanted.

Umbilical Cord   The cord connecting the embryo to the placenta and allowing for the transference of nourishment and oxygen from the mother and of waste material from the embryo.

Underextension   In childhood language development the use of one word to mean one specific thing while not recognizing its application to others (e.g., flower means *only* roses).

Uterus The womb; the organ in which the fertilized ovum implants itself thus beginning embryonic development.

Vernix Caseosa Translated, this means "cheesy varnish" and refers to the oily substance which covers the infant at birth and protects it from infection.

Walking Reflex An infant reflex in which the infant, held under the arms with its bare feet touching a flat surface, will make leg movements that are much like those of well-coordinated walking.

Wechsler An intelligence test originally designed for use with adults and later adapted for use with children.

Zygote Cell resulting from the union of ovum and sperm.

Index

Terman, Lewis, 198, 203
Tertiary circular reactions, 62
Testing play, 155
Tracking, 37
Transductive reasoning, 115

Unconditioned response, 60

Verbal IQ, 199
Vitamins, 105

Vygotsky, L. I., 122–125

Weschler, David, 199
Westinghouse Learning
   Corporation Report, 166
White, Burton, 88, 117
Whorf, Benjamin, 128
World Health Organization, 108

Zygote, 28